# Resilient Recovery from Disasters

Mittul Vahanvati • Elizabeth Maly
Titaya Sararit

# Resilient Recovery from Disasters

## The Long-Term Outcomes of Post-Disaster Housing Reconstruction in India, Thailand and Japan

palgrave
macmillan

Mittul Vahanvati
School of Global, Urban and
Social Studies
RMIT University
Melbourne, Victoria, Australia

Titaya Sararit
Faculty of Architecture
Chiang Mai University
Chiang Mai, Thailand

Elizabeth Maly
International Research Institute of
Disaster Science, Tohoku University
Sendai, Miyagi, Japan

ISBN 978-981-97-8046-4 ISBN 978-981-97-8047-1 (eBook)
https://doi.org/10.1007/978-981-97-8047-1



This Palgrave Macmillan imprint is published by the registered company Springer Nature Singapore Pty Ltd.
The registered company address is: 152 Beach Road, #21-01/04 Gateway East, Singapore 189721, Singapore

If disposing of this product, please recycle the paper.

# Foreword

Post-disaster housing recovery has been described as one of the most intractable challenges in post-disaster recovery. There are many reasons why this is true—the scale of need, costs, the complexity, the associated issues and challenges of housing provision, balancing the meeting of immediate post-disaster needs while investing in long-term homes for living in, and so on.

Learning from experience about how to meet these challenges better is why this book is so important. As the authors set out by saying, the 'new abnormal' is the increase in rapid-onset disasters, many fuelled by climate change. Add to this the complexities of rapid urbanisation, and the result is a greater incidence of disaster, with the greater need in getting post-disaster housing right.

In this book, the authors have nailed their colours to the mast in its very title: resilient recovery. Shaky, partial and uncertain recovery is very often what occurs after disaster, which benefits few and leaves more stranded. Resilient recovery however is about balancing several factors. Building back better is about improving what was before to prevent what might just happened from reoccurring. Resilient recovery means thinking both short and long term—the latter if often forgotten in the rush of aid recovery, but to not consider long-term consequences of short-term decisions is to waste valuable investments and store up future problems. To comment meaningfully on this, the authors visited their case study locations twice, with a ten-year gap between each visit. This rarely happens, and the visits enabled the authors to witness first-hand how early-on decisions played out over a decade in the complexities of real life.

In choosing case studies of disaster recovery in India, Thailand and Japan, the authors are also able to cleverly draw out comparisons between different cultures, societies and systems of governance for what works and what can be improved on. Japan arguably leads the world in disaster prevention and preparedness, yet as elsewhere is still susceptible to overwhelming events. Both India and Thailand have their unfair share of hazards too, and have long experience in what works and what can be improved.

The timing of this book is important. Not only because of increasing need, but because of the challenges to the very essence of post-disaster recovery faced by the humanitarian aid sector. Accepted business-as-usual since the end of World War Two is threatened as global aid funding is being scaled back, at a time when need is increasing. More will need to be done with less, and post-disaster responses need to be more adept and better thought through.

The authors say that, rather than presenting a prescribed manual, they are representing a cookbook with suggestions on recipes. That is an appropriate comparison, as the combination of ingredients needed to suit different cultural and societal tastes is vital for a satisfying outcome. Thankfully the authors avoid suggesting the use of cookie cutters, which have been used by too many insensitive and top-down 'chefs'.

The book concludes with identifying pointers for building long-term resilience post disasters, including building on strengths, collaborative governance, a louder voice for civil society organisations and being wary of top-down approaches. All these are true and vital, and implicit throughout is the need for people's voices to be heard. Some years ago the leading development thinker Robert Chambers coined the phrase 'whose reality counts?' This has never been more true, if we are to adopt the lessons presented in this book to achieve resilient recovery in post-disaster housing.

Sydney, NSW, Australia David Sanderson

# Preface

The idea for this book germinated when Mittul Vahanvati and Liz Maly met at an international conference in Florida in 2019. Our informal discussions on the limited research on long-term outcomes of post-disaster housing recovery projects sparked this collaborative journey, and we invited Titaya Sararit to join our continuing discussions.

If you are reading this book, you are already aware of challenges facing housing at the intersection of disasters and development. Despite adequate housing being enshrined as a fundamental human right by the United Nations, housing has remained a moving target, pointing to a systemic failure of the shared global commitment to meet housing needs for all. Moreover, disasters threaten to damage weak housing of vulnerable people, causing displacement and homelessness. Repeated and frequent cycles of disaster, response, and recovery push the poor and marginalized further down a spiral of poverty, impeding their human rights. Considering these unequal impacts of disasters on housing of the poor, critics argue the "unnaturalness" of disasters.

Climate change will continue to exacerbate hazard intensity, frequency, complexity, and uncertainty. The world experienced one devastating disaster per day over two decades (between 2000 and 2019), and data from 1998 to 2017 suggests an astonishing 91 percent of all hazard events were climate related. Economic losses from disaster rose a staggering 151 percent between the two 20-year periods of 1978–1997 and 1998–2017 (CRED EM-DAT & UNISDR, 2018).

A "new abnormal" created by the changing climate demands innovative ways of planning, housing design, construction, maintenance, and governance. It also encourages systems thinking, promoted by the resilience concept, introduced in disaster studies in 1970s. The resilience concept is promoted as a key objective by the Sendai Framework for Disaster Risk Reduction (UNDRR, 2022), as well as by the Intergovernmental Panel on Climate Change (IPCC, WGII, 2022), in efforts to tackle climate change and create sustainable development pathways. Resilience is defined as:

> the ability of a system, community, or society exposed to hazards to resist, absorb, accommodate, adapt to, transform, and recover from the effects of a hazard in a timely and efficient manner, including through the preservation and restoration of its essential basic structures and functions through risk management. (UNDRR, 2024)

As suggested in this definition, resilience is not just about "bouncing back" to pre-disaster levels of housing but rather about continual adaptation of residents and the housing sector in response to rising risks from hazards and climate change. However, operationalizing the concept of resilience within the housing sector presents unique challenges due to the diversity of interpretations and the complexity of systems thinking.

Several disaster scholars and development practitioners have identified post-disaster housing reconstruction as an opportune time to bring about systemic changes for resilience outcomes. However, it is unclear how to achieve such change. The last 100 years of scholarship on housing, disaster, and development suggests two key barriers to systemic changes from post-disaster housing reconstruction, as (i) resilience narrowly defined as achieving structurally strong housing (or engineering resilience) rather than housing as a system, within a broader system whereby housing rights and resilience are equally important for residents; and (ii) a lack of focus on long-term outcomes of the rebuilding process.

Considering these complexities, we focus on resilience—of housing systems, and—against all disruptions (anticipated and unexpected, natural, or climate related). A housing system is defined as more than a mere physical or economic asset, including *housing as a system* (including services, energy flows, material flows), *and housing embedded within a broader system* (legal planning system and building regulations, culture of compliance to building standards, construction labour skills, enforcement, etc.). During post-disaster housing reconstruction, systemic changes in the housing system can include purposeful realignment of town- or settlement-planning, housing design, and skills in the construction sector, as well as social and political changes to reduce future hazard risk. In regards to timeframes, scholars noted that although resilience building is a continuous process, evident pre-disaster, during disaster, and long-term after disaster (as adaptive resilience), radical or systemic changes are most evident a long time (>10 years) after disaster. However, such long-term evidence is limited.

Some recent scholarship has investigated long-term impacts of housing reconstruction projects (Ahmed et al., 2020, 2023; Barenstein et al., 2014; Barenstein & Leemann, 2012; Davis & Alexander, 2016; Haigh & Amaratunga, 2011; Lizarralde et al., 2010, 2015; Lyons et al., 2010; and Schilderman & Parker, 2014). With trickling research progress on long-term impacts of housing reconstruction, this topic deserves more attention, especially from a systems perspective. To address this gap, we seek answers to the question:

> What are the key factors that catalyzed long-term resilience outcomes after post-disaster housing reconstruction policies and projects?

In this book, we aim to identify key factors that helped or hindered systemic change in the housing system (house, individuals, planning and construction sectors, institutions) towards resilience (Fig. 1), while also upholding housing as a human right.

Due to clear synergies in our research, we the authors—Mittul Vahanvati, Liz Maly, and Titaya Sararit—having examined housing reconstruction and long-term outcomes in three Asian countries—India, Japan, and Thailand—respectively, with the objective of capturing lessons learned. Since we all were trained in architecture yet found our paths into academia, we do not claim to have expertise in human geography, economics, or practical disaster recovery—fields that are undoubtedly crucial to a complete systems-based analysis of post-disaster recovery. However, we have all dedicated over a decade to building our research skills in incorporating systems thinking and examining enduring effects of post-disaster housing reconstruction within our respective regions—Mittul in India, Liz in Japan, and Titaya in Thailand—potentially bringing objectivity and a broader relevance to our research. Being based in three different Asian countries allows us to bring rich cultural insights rooted in our distinctive geographical contexts. This confluence of personal and scholarly experience enriches our inquiry with deep cultural sensitivity. As three female scholars, we also hope to include a gendered perspective on care and capabilities in disaster recovery processes.

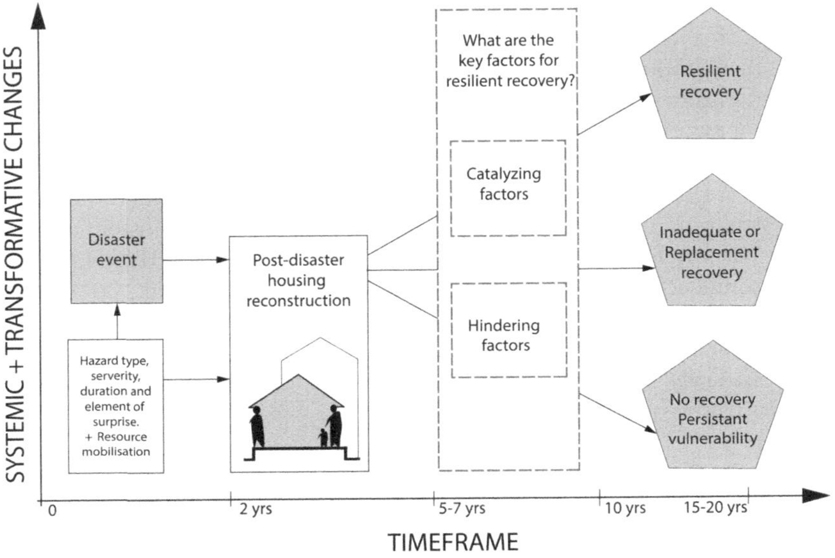


**Fig. 1** Conceptual model on gaps in evidence on factors that link post-disaster housing reconstruction to resilience outcomes in long-term. (Source: Vahanvati, M. with feedback from the authors)

We developed the research method collaboratively and adopted a predominantly qualitative and interpretive methodology, allowing us to have a deeper understanding of contextual characteristics of the housing practice and people's views, and draw relationships between long-term resilience outcomes and key factors of housing reconstruction (actors, activities, resources, speed, and duration), which instigated those outcomes, as per the theory of change.

Since Asian countries account for 80 percent of global disaster deaths and 76 percent of affected people, despite experiencing 45 percent of disaster events (CRED, 2018), research on Asia felt crucial. Moreover, data suggests that low- to middle-income countries were disproportionately affected by the greatest proportion (68 percent) of deaths over a 20-year (1998–2018) period, despite having experienced only 43 percent of all global disasters (CRED EM-DAT & UNISDR, 2018. p. 17). In addition, over the two decades mentioned earlier, disasters caused seven times more deaths in low-income communities than their high-income counterparts (Iyer-Raniga & Vahanvati, 2021). Asia's large landmass, high relative population density, low-middle-income levels, and multiple/complex hazard risks contribute to such disproportionate impacts. This evidence encouraged us to focus on three Asian countries at different development stages (e.g. India–lower-middle income, Thailand–upper-middle income, and Japan–high income). We acknowledge that a country's income level or gross domestic product (GDP) is not the only influencing factor for disaster impact or recovery ability. Other factors at play include a country's governance and societal conditions and potentially, Human Development Index (HDI). The three selected Asian countries not only present diverse economic statuses, they also have diverse social, cultural, and political contexts, varied hazard exposure, vulnerabilities, and disaster risk management strategies, offering a rich basis and fresh geographic scope for a whole-system perspective and synthesis of learning for potential applications across wider geographies.

For long-term cross-country and cross-cultural analysis, we selected two cases—post-disaster housing reconstruction projects from each country, one from 24–30 years ago and another from 14–17 years ago. Selected cases were communities affected by the 2001 Gujarat Earthquake and the 2008 Kosi River floods in India, the 2004 Indian Ocean Tsunami and the 2011 Floods in Ayutthaya in Thailand, and the 1995 Great Hanshin Awaji (Kobe) Earthquake and 2011 Great East Japan Earthquake and Tsunami in Japan. Criteria for the selection of case studies included: (i) the authors' long-term research in those countries; (ii) practicalities of data collection for longitudinal, cross-country, multi-dimensional analysis; and (iii) the diverse social, cultural, political, and economic contexts of each case and country (Fig. 2, Table 1).

Each author has conducted field visits in their respective country, twice—once a decade ago (2013–2015) and again, in recent years (2022–2024). We visually assessed rebuilt houses using photographs and sketches to evaluate the construction quality and modifications over time. We collected qualitative data through semi-structured interviews and/or focus groups with three types of stakeholders: (1) disaster-affected community members and neighborhood leaders, (2) stakeholders such as government, non-government organizations (NGOs), civil society organizations (CSOs), and non-profit organizations, and (3) built environment professionals, including community planners and local artisans. Data was gathered to understand community and institutional or professional perceptions, knowledge, and abilities, and corroborate with our observed evidence of rebuilt house quality, house extension quality,

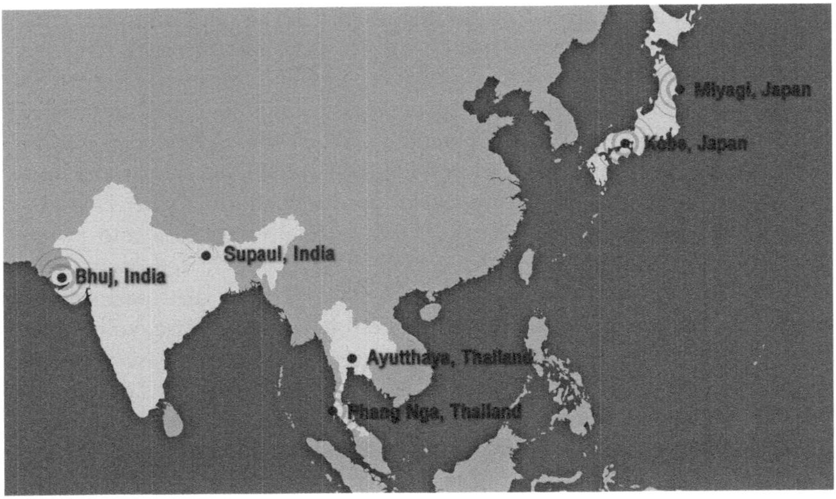


**Fig. 2** Map of case study locations in India, Thailand and Japan (Source: Authors)

replication of safe construction techniques, activities on the street, overall living standards of residents, and feeling of the neighborhood. The empirical research in each country followed human subjects' protocols, with approval for this research granted by ethics committees from the authors' three respective universities: The Royal Melbourne Institute of Technology in Australia (25 October 2022/No. 25853), Chiang Mai University in Thailand (25 April 2023/No. CMUREC 66/067), and Tohoku University IRIDeS in Japan (28 November 2022/No. 2022-066). Our empirical data was complimented by secondary data, including reference materials, industry reports, and recovery literature.

For data analysis, we considered the following four time periods to examine the longer-term outcomes of post-disaster housing reconstruction in case study sites:

1. Pre-disaster housing context,
2. Disaster impacts and housing reconstruction policies and programs,
3. Post-disaster housing reconstruction implementation and results, and
4. Long-term outcomes more than 10 years post disaster.

**Table 1** Overview of two disasters from each country: India, Thailand and Japan, for cross-cultural analysis (Source: Authors, from various sources)

| *Disaster* | *2001 Gujarat Earthquake (EQ), Bhuj, India* | *2008 Kosi River floods, Bihar, India* | *2004 Indian Ocean Tsunami, Namkem, Thailand* | *2011 Ayutthaya Flood, Huntra, Thailand* | *1995 Great Hanshin Awaji Earthquake (EQ), Kobe, Japan* | *2011 Great East Japan Eurthquake (GEJE) and Tsunami, Shichigahama, Japan* |
|---|---|---|---|---|---|---|
| Hazard magnitude | 7.9 magnitude | n/a | 8.9 Richter scale | n/a | 7.2 magnitude | 9.0 magnitude |
| Deaths or displaced | 20K deaths | 3 million people displaced | 230K deaths | 813 deaths | ~6.4K deaths | 19K+ deaths 350K people displaced |
| Houses destroyed or damaged | 1 million+ houses destroyed/ damaged | 236K+ houses damaged | Unknown | 1.96 million damaged 19,000 (1%) destroyed | 240K houses destroyed | 122K houses destroyed |
| Economic loss | 3.5 bullion USD | 195 million USD | Unknown | Unknown | $ 99 billion USD | ¥16.9 trillion ($210 billion USD) |
| Duration of housing reconstruction | 2+ years | 10 years | 2+ years | 3 years | 5 years | 10 years |

Long-term outcomes of housing reconstruction projects were analyzed against five resilience characteristics: (i) Robustness and diversity, (ii) Appropriateness and adaptability, (iii) Redundancy and integration, (iv) Rights, equity, and security, and (v) Resourcefulness (or 4R+A), across all case studies. Following the analysis of similarities, differences, and changes over time between the two case studies in each country, we looked back to relate outcomes across national boundaries to the reconstruction processes.

## Structure of the Book

Chapter 1 takes a retrospective journey to scrutinize waves of international reforms in the housing sector, and their intersection with disaster studies and dominant development paradigms since the late eighteenth century, to understand how we got here—power imbalances, social inequity, disaster vulnerability, and environmental damage. This critique highlights putting the most vulnerable people at the center of decision-making for their own housing. The chapter ends with key definitions and a conceptual framework for resilience characteristics. Chapters 2, 3, and 4 dive into deeper analysis of two case studies from India, Thailand, and Japan, respectively. Chapter 5 extends the conversation beyond findings from individual countries to a cross-country and cross-cultural synthesis. This chapter distills similarities, differences, and observable shifts in policies and practice that bear on enhancing long-term resilience in the housing system. Critical analysis of empirical evidence leads to suggestions about effective/catalytic strategies for resilient housing recovery. Key recommendations, informed by identified strategies, are put forward with the hope to inform future housing reconstruction interventions and not repeat past mistakes.

## Who Is This Book for?

This book is a call to action for recovery policy makers and practitioners to leverage future foresight and planning capacities, pre-disaster, to achieve long-term resilience. Recommendations gathered from experiences in Asian countries suggest that if these places with limited resources and massive developmental challenges can leapfrog towards resilient and sustainable pathways, lessons from these experiences can be applicable for future recovery interventions, regionally and globally.

This volume is not a manual providing prescriptive steps for on-the-ground recovery operations. Rather, it is like a cookbook that offers recipes, adaptable to a cook's taste and available ingredients, providing 'key principles' for resilient housing—spanning recovery programming, planning, design, construction, supervision, and maintenance—that collectively deliver lasting benefits for communities and the environment. These key principles are intended to be adapted to meet the varied and complex needs of specific locales and societies. We steer clear of probing the origins of hazards, whether natural, anthropogenic, or exacerbated by climate change. Instead, our focus lies on unpacking the multifaceted recovery following the impacts of these hazards and understanding how they interplay with societal vulnerabilities and capacities. Our aim is to uncover strategies that mitigate the compounding effects of such hazards and contribute positively to the long-term trajectories of housing recovery to resilience.

Melbourne, VIC, Australia — Mittul Vahanvati
Sendai, Japan — Elizabeth Maly
Chiang Mai, Thailand — Titaya Sararit

## Note

1. Hazards are not same as disasters, and disasters are not natural. While hazards may have natural occurrence (e.g. geological plate shifting causing earthquakes), they become disasters only when they interact with society that is not well prepared. For example, an earthquake in a wasteland is not called a disaster because it has not led to any loss of life or damage to properties. Therefore, disasters, can be natural hazard- or climate-induced; however, they are never 'natural' disasters, as disasters are socio-politically constructed.
2. Recovery vs Reconstruction: The term recovery is used to describe holistic restoration/improvement in people's health, livelihood, economic, physical, social, cultural and environment assets, whereas reconstruction is used to describe medium- and long-term rebuilding of physical assets such as housing, infrastructure, or services.
3. Resilient recovery vs Recovery is discussed further in Chap. 1.

## References

Ahmed, I. N., Masa, O'Brien, D., Tucker, C., & Vahanvati, M. (2023). *Resilience and Sustainability in Architecture and Urban Planning: Policies, Practices, Strategies and Visions.* MDPI—Multidisciplinary Digital Publishing Institute. https://doi.org/10.3390/books978-3-0365-6418-0

Ahmed, K. I. a., Maund, K., & Gajendran, T. (2020). *Disaster Resilience in South Asia: Tackling the Odds in the Sub-continental Fringes*. Routledge.

Barenstein, J. D., Modan, A. N., Talha, K., Upadhyay, N., & Khandhadai, C. (2014). Looking Back at Agency-Driven Housing Reconstruction in India: Case Studies from Maharashtra, Gujarat and Tamil Nadu. In T. Schilderman & E. Parker (Eds.), *Still Standing? Looking Back at Reconstruction and Disaster Risk Reduction in Housing* (pp. 39–58). Practical Action.

Barenstein, J. E., & Leemann, E. (2012). *Post-disaster Reconstruction and Change: Communities' Perspectives*. CRC Press.

CRED EM-DAT, & UNISDR. (2018). *Economic Losses, Poverty & Disasters 1998–2017* CREDCrunch. Centre for Research on the Epidemiology of Disaster. Retrieved from https://www.undrr.org/publication/economic-losses-poverty-disasters-1998-2017#:~:text=In%201998%2D2017%2C%20disaster%2D,reported%20between%201978%20and%201997

Davis, I., & Alexander, D. (2016). Recovery from Disaster. Taylor and Francis. https://doi.org/10.4324/9781315679808

Haigh, R., & Amaratunga, D. (2011). *Post-disaster Reconstruction of the Built Environment Rebuilding for Resilience*. Wiley-Blackwell. Retrieved from http://site.ebrary.com/lib/rmitau/Doc?id=10483295

Iyer-Raniga, U., & Vahanvati, M. (2021). Resilience of the Poor and Vulnerable Against Disasters and Associated Economic Shock. In W. Leal Filho, A. M. Azul, L. Brandli, A. Lange Salvia, P. G. Özuyar, & T. Wall (Eds.), *No Poverty: Encyclopedia of the UN Sustainable Development Goals* (1st ed., pp. 791–803). Springer Nature. https://doi.org/10.1007/978-3-319-95714-2_59

Lizarralde, G., Johnson, C., & Davidson, C. (2010). From Complexity to Strategic Planning for Sustainable Reconstruction. In *Rebuilding after Disasters from Emergency to Sustainability* (pp. 248–225). Spon Press.

Lizarralde, G., Valladares, A., Olivera, A., Bornstein, L., Gould, K., & Barenstein, J. D. (2015). A Systems Approach to Resilience in the Built Environment: The Case of Cuba. *Disasters*, *39*(s1), s76–s95. https://doi.org/10.1111/disa.12109

Lyons, M., Schilderman, T., & Boano, C. (2010). *Building Back Better: Delivering People-Centred Housing Reconstruction at Scale*. Practical Action.

Schilderman, T., & Parker, E. (2014). *Still Standing? Looking Back at Reconstruction and Disaster Risk Reduction in Housing*. Practical Action.

UNDRR. (2022). *Global Assessment Report on Disaster Risk Reduction: Our World at Risk: Transforming Governance for a Resilient Future*. UNDRR. Retrieved from https://www.undrr.org/gar2022-our-world-risk

# Acknowledgements

This research would not have been possible without the generosity of all the disaster survivors and disaster recovery stakeholders who shared their precious time and reflections with us, for which we are humbled and extremely grateful. We deeply appreciate the support we received for the research that became this book, from the Earthquake Research Center of Thailand (EARTH), Dr. Haruka Tsukuda, Mr. Shigeki Ogino, and Dr. Masato Tanaka. We would also like to express our gratitude for support received from Dr. Karyn Bosomworth, Prof. Usha Iyer-Raniga, Prof. Darryn McEvoy, Dr. Yoshimitsu Shiozaki, Dr. Tamiyo Kondo, Prof. David Sanderson and Associate Prof. Iftekhar Ahmed, for believing in us and supporting us in our academic journey. We are also thankful for the help of Aom. Ominrada Bootkaew for helping create the map images. Lastly, we would like to thank our family members-Munir Vahanvati, Meer Vahanvati, Kabir Vahanvati, Jamila Vahanvati, and Urmila Shah, without whose unconditional support and motivation, and taking off the burden of household chores enabling us to complete this mammoth task of writing a book.

# Contents

# About the Authors

**Mittul Vahanvati** is a program manager and a senior lecturer in the Sustainability and Urban Planning program at RMIT University in Melbourne, Australia. Her research focuses on the long-term impacts of housing reconstruction projects and recovery programmes, for community resilience. Her research aims to build capacities of households, professionals and policy makers to make risk-informed decisions for disaster resilience and climate adaptation. Trained as an architect and practiced in architecture industry for over 10 years, she is also a co-founder of a Melbourne-based design+build studio, Giant Grass. She adopts co-design and design-thinking approaches in her applied research, in the context of India, Solomon Islands, and Australia. ORCID ID: https://orcid.org/0000-0003-3336-0937

**Elizabeth Maly** is an associate professor at the International Research Institute of Disaster Science, Tohoku University, in Sendai, Japan. With the theme of people-centered housing recovery, her research interests are community-based housing recovery and temporary, transitional, and permanent housing provision within reconstruction—including policy, process, and housing form—that support successful life recovery for disaster-affected people. Past and current research focuses on the experiences of people affected by disaster and the roles of government and NGOs in the processes of housing reconstruction and resettlement after disasters in the USA, Indonesia, Philippines, and Japan. ORCID ID: https://orcid.org/0000-0003-2827-1278

**Titaya Sararit** is an assistant professor and Deputy Dean for Academic Affairs & Research at the Faculty of Architecture, Chiang Mai University, in Chiang Mai, Thailand. She has a PhD in Architecture from Kobe University, Japan. Her research interests are housing recovery, temporary shelter for local community, community-based design, and long-term recovery after flood, earthquake, and tsunami in Thailand. ORCID ID: https://orcid.org/0000-0002-0239-2402

# Abbreviations

| | |
|---|---|
| ADB | The Asian Development Bank |
| BAPEPS | Bihar *Aapada Punarwas Evam Punarnirman Society* |
| BHADA | Bhuj Area Development Authority, India |
| CAPAM | Commonwealth Association for Public Administration and Management |
| CGI | Corrugated Galvanised Iron Sheet |
| CODI | Community Organization Development Institute |
| COP | United Nations Climate Conference of the Parties |
| CRED | Centre for Research on the Epidemiology of Disasters |
| CSOs | Civil Society Organizations |
| EBC | Extremely Backward Class |
| EPC | Environmental Planning Collaborative |
| EWS | Extremely Weaker Social group |
| GDP | Gross Development Product |
| GEJE | Great East Japan Earthquake |
| GFDRR | Global Facility for Disaster Reduction and Recovery |
| GIDC | Gujarat Industrial Development Corporation |
| GoB | Government of Bihar |
| GoI | Government of India |
| GEERP | The Gujarat Emergency Earthquake Reconstruction Project |
| GSDMA | The Gujarat State Disaster Management Authority |
| HDFC | Housing Development Finance Corporation |
| IFRC | International Federation of Red Cross and Red Crescent Societies |
| IPCC | The Intergovernmental Panel on Climate Change |
| KFW | Kreditanstalt für Wiederaufbau (translated as Credit Institute for Reconstruction) |
| KMVS | Kachchh Mahila Vikas Sangathan |

| | |
|---|---|
| KRIC | The Kachchh Rehabilitation Information Centre |
| KSKs | Kosi Setu Kendras (translated as Kosi River Bridging Hubs) |
| NATO | North Atlantic Treaty Organization |
| NDMA | National Disaster Management Authority |
| NPO | Non-Profit Organisations |
| OBC | Other Backward Classes |
| ODA | Official Development Assistance |
| ODHR | Owner-Driven Housing Reconstruction |
| OECD | The Organization for Economic Cooperation and Development |
| OHCHR | Office of the High Commissioner for Human Rights |
| ODRC | Owner-Driven Reconstruction Collaborative |
| PMAY | Pradhan Mantry Awas Yogna (translated as Prime Minister's Housing Programme) |
| RCC | Reinforced Cement Concrete |
| SFDRR | Sendai Framework for Disaster Risk Reduction |
| SDGs | Sustainable Development Goals |
| SC | Scheduled Caste |
| ST | Scheduled Tribe |
| UN-Habitat | United Nations Human Settlements Programme |
| UN-DESA | United Nations Department of Economic and Social Affairs |
| UNDP | United Nations Development Programme |
| UNDRO | United Nations Disaster Relief Organization |
| UNDRR | United Nations Office for Disaster Risk Reduction (previously known as UNISDR) |
| UNFCCC | United Nations Framework Convention on Climate Change |
| UNISDR | United Nations International Strategy for Disaster Reduction (now UNDRR) |
| USAID | U.S. Agency for International Development |
| WMO | World Meteorological Organization |

# List of Figures

# List of Tables

CHAPTER 1

# A Changing Development Geography Intertwined with Disasters, Housing, Recovery, and Resilience

**Abstract** This chapter focuses on post-disaster housing reconstruction and its link to rights, justice, and resilience due to society's pursuit of a "dominant" development model. It draws on scholarly literature in disasters, development, and built environment studies since the late eighteenth century. This critique shows that the dominant growth-driven development model is unsustainable as it has perpetuated power imbalances, marginalized the vulnerable, perpetuated underdevelopment, and exploited the environment, as evident by climate change. Continuing on the current development path is unlikely to yield the changes essential for human society to survive in a rapidly changing climate. Post-disaster housing recovery offers an opportunity for a systemic and radical change. The chapter calls for a radical shift towards post-development models of housing reconstruction that encourages localized, place-based, environmentally sensitive, empowering, and people-centered approaches, particularly for marginalized communities in the Asia-Pacific region. It advocates for a systems approach, highlighted in the concept of resilience. While participatory approaches are well-researched, the concept of resilience lacks clarity, or a normative dimension and operationalizability during housing recovery. The chapter contributes to discourses linking post-disaster housing reconstruction to long term resilience by distilling essential attributes of resilient recovery.



M. Vahanvati et al., *Resilient Recovery from Disasters*,
https://doi.org/10.1007/978-981-97-8047-1_1

**Keywords** Disasters • Development • Housing • Adequate housing • Housing reconstruction • Rights • Resilience • Vulnerability • Exposure • Disaster resilience • Disaster risk reduction • Build back better • People-centered • Owner-driven • Localization • Long-term recovery • Resilient recovery • Recovery • Robustness • Adaptive capacities

> *"Nothing undermines development like a disaster."* (UNDRR, 2022)

Understanding of the term "development" and its relationship to disasters has evolved over time. Development is understood as improvement in social, economic, and political conditions; disasters are considered as disrupting this development process. This chapter critiques how the term "development" and its progression, practice, and outcomes have been predominantly determined by Western-, Anglo-, or Euro-centric approaches and imposed upon the developing world (including many former colonies). Impacts of such exploitative practices are evident in ecological damage (e.g. climate change) and exacerbated social inequities. Sachs (2010, p. VI) aptly notes that

> *"The desire for equity is largely fixed on development-as-growth, and it is development-as-growth, which strains human relations and fundamentally threatens the biosphere."*

In this chapter, we critically explore how housing reconstruction interventions, in the context of the Western-dominant development model, have dominated since the eighteenth century. These processes have perpetuated power imbalances between rich and poor countries, marginalized the vulnerable, and in some cases diminished their autonomy. Post-development theory emerged as a response urging for localized development models that returned agency, control, and freedom of decision-making to the local people. We consider the current state of knowledge in disaster studies in this context, which requires an understanding of key concepts—vulnerability, exposure, risk, and resilience—all which advocate for systemic change. This chapter's conclusion outlines key characteristics of the concept of resilience and its relationship to housing to examine if and how it can be operationalized in long-term recovery. Interwoven throughout this chapter are ideas of how housing recovery might carve pathways towards resilient recovery and resilient development, which can adapt to the changing climate.

## Disasters and Development

Historically considered completely separate from development, demonstrated by fatalistic views, disasters were expressed as God's rage, or "erroneous meteorological" or "terrible" events (Blasius, 1877, p. 215) and perceived to be a "natural" occurrence. However, since the 1970s, there has been a gradual shift towards scientific understanding of causes and effects of disasters. This is evident in the formation in 1988 of the Centre for Research on the Epidemiology of Disasters (CRED EM-DAT, 2009), which compiles a global disaster database by hazard type and hazard impacts in terms of fatalities and people affected.

## Genesis and Evolution of the Term 'Development'

The term "development" was first used in a modern context by President Harry Truman of the United States in the 1940s, signifying the pursuit of poverty eradication through economic growth (Sachs, 2010). Behind this ostensibly noble quest lay a broader geopolitical motive to redistribute global power, particularly from Western Europe to the new superpowers of the time, the United States and the Soviet Union (Oliver-Smith, 2010). Subsequent decades saw "development" become synonymous with capital- and carbon-intensive growth (Cambridge Dictionary, 2023), often neglecting environmental, cultural, and social considerations.

Despite early warnings about the environmental harm caused by industrial practices, notably the use of coal, the appeal of progress as epitomized by the industrialization-fueled improvement of living standards overruled concerns. Evidence of environmental harm from fossil fuels has existed since the 1820s, and in the 1890s Svante Arrhenius, a Swedish physicist, chemist, and Nobel Prize winner (Grant, 2019), compiled evidence proving human influence on climate change. However, society's pursuit of economic growth spurred industrialization, modernizing living conditions, health, education, and infrastructure.

The post-World War II Marshall Plan solidified this ideal. The Marshall Plan was the first large-scale housing reconstruction program, funded by the U.S. from 1948 to 1952, with the aim of rebuilding housing, infrastructure, and economy in war-ravaged European nations. It promoted mass production of a uniform, modern house, using industrial materials, as a symbol of recovery. A one-size-fits-all approach, disregarding diverse socio-cultural and climatic needs across populations (Sachs, 2010) was justified in the name of equality, speed, and resource efficiency (Lizarralde et al., 2010). With a strong emphasis on top-down, industry-derived solutions, the Marshall Plan and other post-war recovery strategies laid the foundation for standardizing housing reconstruction.

## Housing Recovery as Foreign Aid, During Globalization

This period also saw the advent of foreign aid as a cornerstone of development policy—potentially strengthening ties between donor and recipient nations, yet sometimes reinforcing dependency and wealth concentration rather than facilitating genuine socioeconomic empowerment. Foreign aid distribution from developed to low-income countries became a key tenet of development or post-disaster housing recovery. For example, the U.S. drove the establishment of several aid and welfare organizations, including the World Bank groups and NATO. Later, Western European countries, Canada, Japan, Australia, and New Zealand followed the U.S. by joining the Organization for Economic Co-operation and Development (OECD). Combined, these nations provide most Official Development Assistance (ODA) to developing countries.

Conferences like Habitat I, in 1976 in Vancouver, Canada, galvanized this approach further, advocating for developed countries to finance and provide technical support for housing recovery aimed at providing "adequate shelter for all" (UN-Habitat, 1976).

"Adequate housing," a fundamental human right stated in Article 25(1) of the Universal Declaration of Human Rights (1948) (OHCHR & UN-Habitat, 2009), is defined as: (i) robust or safe to withstand future hazards to provide protection to its residents; (ii) secure in terms of land rights and tenure; (iii) well located as per the livelihood or other preferences of its residents; (iv) culturally adequate; (v) service integrated for access to safe drinking water, sanitation, and energy; (vi) affordable; and (vii) accessible or inclusive for all (OHCHR & UN-Habitat, 2009). A multifaceted asset, adequate housing plays a crucial role in improving residents' overall quality of life, sense of place, dignity, health, education, socialization, safety, and economic prospects (Barakat, 2003; OHCHR &

UN-Habitat, 2009) as well as freedom from discrimination or inhuman treatment. Yet, "adequate" became a moving target, as the rapidly urbanizing world grappled with swelling numbers of slum dwellings and unceasing housing crises, pointing to a systemic failure of the shared global commitment to housing as a human right.

## A Call for Participatory Housing Recovery Amid Growing Concerns of Aid Dependency

Witnessing the inadequacies of a purely aid-driven recovery, scholars and practitioners argued for a reimagined approach that places affected communities at the heart of housing reconstruction. Critics, particularly from Latin America, noted that Third World poverty did not just happen, but rather it was produced from centuries of colonization and continued exploitation of developing countries by developed ones, burdening them with interest on loans and massive debt repayments (Warf, 2001). This historic trajectory gave rise to the Dependency Theory, which posited that modernization efforts, as seen in aid-driven economic growth agendas, had often perpetuated exploitative relationships rather than ameliorated poverty.

Marking a significant paradigm shift away from top-down and aid-driven models, thought leaders called for participatory and self-help approaches where community capabilities are intrinsic in housing recovery. These thought leaders include Robert Chambers (1983); Ian Davis (1978), who studied the aftermath of the 1976 earthquake in Guatemala and Mexico; Peter Winchester (1979), who worked in the aftermath of the 1977 cyclone in Andhra Pradesh in India; John Turner (1976), who observed informal settlement in Peru; Sultan Barakat (2003); and U.S.-based anthropologist Anthony Oliver-Smith (1996), who examined the 1970s avalanche and earthquake in Peru. They emphasized that more than a mere a physical asset, a house was also a social asset integral to cultural expressions of people; therefore houses are best built with the involvements of residents, through mutual help practices or self-help and culturally rooted practices, rather than top-down interventions. These practices are not new; they were prevalent globally in traditional or indigenous cultures, including the *Ayni practice* in rural settlements in the Andean region of Latin America (Barakat, 2003). The first comprehensive guideline on *Shelter after disaster* (UNDRO, 1982, 2010) validated and incorporated these assertions, and Davis and Alexander (2016) and Jha et al. (2010) meticulously consolidated the evolution in thought and practice about these recovery and reconstruction approaches and objectives.

Reflecting a greater emphasis on empowerment and sustainability in recovery practices, the Owner-Driven Housing Reconstruction (ODHR) approach crystallized from this philosophy and was mainstreamed, evident by the *ODHR Guidelines* by the International Federation of Red Cross and Red Crescent Societies (IFRC, 2010) and *the Handbook on Owner-Driven Housing Reconstruction* by UNDP (2022). Early adoption of ODHR was prompted after disasters including the 2001 Gujarat Earthquake in India, 2004 Indian Ocean Tsunami, 2005 earthquake in Pakistan, and 2010 Haiti earthquakes. In contrast to top-down or donor-driven approaches, ODHR stands out for its speed, cost-effectiveness, and empowerment (Barenstein & Iyengar, 2010; Hunnarshala, 2007; Vahanvati, 2018b), and most importantly, it puts people—their values, needs, aspirations, and capacities—at the center of the recovery process.

However, challenges persist in reconciling development aspirations of poverty-stricken communities, who despite culturally and climatically inappropriate housing (Barenstein & Leemann, 2012), have conflicting views about modern houses as the symbol of development and a way out of poverty (Sachs, 2010) rather than a cause of impoverishment.

## Post-Development and Sustainable Development

The late 1990s to early 2000s heralded the arrival of the post-development era, driven by heightened social and environment consciousness. The discourse around development began to challenge its traditional trajectory, turning towards sustainable and human-centered models of growth. This ideological shift recognized the limitations of industrialization and sought to prioritize human and ecological well-being over unchecked economic expansion.

Urbanization trends have not reduced the prevalence of inadequate housing. Additionally, Habitat II in 1996 in Istanbul, Turkey, identified that developed countries were not meeting the UN-mandated 0.7 percent of their GDP to be used for financial aid (Roser, 2021; UN-Habitat, 2003). The *Challenge of Slums* report by UN-Habitat (2003, p. XXV–XXVI) noted

> *In 2001, 924 million people, or 31.6 percent of the world's urban population, lived in slums. The majority were in developing regions, accounting for 43 percent of the urban population, in contrast to 6 percent in more developed regions .... It is almost certain that slum dwellers increased substantially during the 1990s. It is further projected that in the next 30 years, the global number of slum dwellers will increase to about 2 billion if no firm and concrete action is taken.*

This report forecasted a doubling of the current housing crisis by 2030, providing a stark reminder of the work that remains to be done.

As the concept of development matured, it drew increasingly from various disciplines to redefine progress. Concerns over social inequities prompted economists like E.F. Schumacher (1993) in *Small is Beautiful* (1973) and Amartya Sen (1985) with his human capability approach to advocate for a more humane measure of development—Human Development—that prioritized individual capabilities and community assets over gross economic output. Complementary propositions of U.K.-based academic and practitioner Robert Chambers (1995) highlighted the importance of "sustainable livelihoods" as a pathway to achieving adequate housing and overall well-being.

This reconceptualization of development coincided with growing environmental consciousness. Rachel Carson's (2002) landmark work *Silent Spring* served as a clarion call in 1962, providing undeniable evidence of the ecological costs of economic expansion. Ensuing decades witnessed a growing body of research that laid bare the ramifications of unchecked development—declining biodiversity, contaminated water supplies, and altered climate patterns. Examples include reports indicating growing genetic mutations among birds, use of pesticide leading to a 50% decrease in male sperm counts since 1938, increases in toxins in human breast milk (Orr, 2004), and rising levels of carbon dioxide from 280 parts per million (ppm) over the last 12 million years (Papanek, 1995, p. 22) to 410 ppm in 2019 (IPCC, 2023). The "Great Acceleration" in carbon emissions following the industrial revolution—a critical turning point highlighted in the latest IPCC (2022) report—can be seen as a collective acknowledgment of human impacts on climate change, termed the Anthropocene, which has been recognized but insufficiently addressed over the past century.

The journey of "development" as a term and its goals reflect an evolution in global priorities—from the foundational Millennium Development

Goals (MDGs) focused on social, environmental, and economic objectives, to the holistic approach of the Sustainable Development Goals (SDGs) focusing on people, planet, prosperity, peace, and partnerships. The SDGs comprise 17 interconnected goals, forming part of a global blueprint adopted by the UN and designed to foster a sustainable future that addresses intergenerational needs amid an ever more apparent climate crisis. While these goals exemplify the global consensus on the need for sustainable development, the on-the-ground reality points to a slow and uneven progress, emphasizing the urgent necessity for transformative actions in alignment with the finite nature of Earth's resources.

Despite the post-development era's focus on sustainable and human-centered growth, fascination with rapid economic progress continued to dominate global priorities. Rather than departing from growth-oriented strategies, society increasingly leaned on market-driven mechanisms as a harbinger of globalization—a movement marked by the liberalization of trade under the aegis of the World Trade Organization (founded in 1995, previously termed the General Agreement on Tariffs and Trade, formed in 1948) (Klein, 2014). While globalization spurred economic development, lifting a significant portion of the global population out of poverty, it simultaneously failed to address the persistent state of poverty (Roser, 2021), in which the majority (85% in 2021) of the world's population still lives.

Habitat III in 2016 brought the housing sector, particularly affected by global poverty, into sharp focus (United Nations General Assembly, 2017), underscoring challenges of a rapidly urbanizing world and the proliferation of informal settlements. It marked the conundrum of modernity, with its triad of industrialization, capitalism, and economic growth, which has fallen short in fulfilling the universal right to housing. In 2023, two billion people faced homelessness or lived in conditions deemed inadequate—a crisis projected to escalate to three billion (40 percent of the world's population) by 2030, if current development trajectories remain unchanged (UN-Habitat, 2022, 2023). Addressing the global housing crisis requires approximately 96,000 new affordable homes be built daily until 2030 (UN-Habitat, 2023).

The persistence of the housing crisis and risk it poses to safe living conditions highlights the limitations of relying solely on foreign aid and the global free trade paradigm, which has often resulted in exacerbating wealth disparities and creating neo-colonial dependencies rather than fostering equitable resource distribution. To respond to the burgeoning housing deficit and ensure disaster resilience, a monumental effort is required—one that aligns with the spirit of post-development thinking, advocates for systemic change, and calls for prioritizing the provision of affordable, adequate, and resilient housing on a global scale.

## Systems Thinking: Disaster Risk and "Build Back Better"

From traditional to contemporary ideologies, the shift in development discourse continued with the emergence of systems thinking, which reconceptualized disaster risk as a relational dynamic among hazard, exposure, vulnerability, and capacity. Systems thinking challenges the presumption of the "naturalness" of disasters, spotlighting how societal structures and power relations intensify impacts of hazards.

Rising in prominence as a response to the complexity of socio-ecological challenges, systems thinking inspired scholars from varied disciplines (Wisner et al., 2004) to devise a more layered understanding of disaster risk. This view, endorsed by UNDRR (2024), posits risk as a function of multiple intertwined factors: hazard x exposure x vulnerability (vulnerability = sensitivities/capacities)—each deeply ingrained in the societal matrix.

The influential concept of disaster risk was later defined by UNDRR (2024) as:

> *The potential loss of life, injury, or destroyed or damaged assets which could occur to a system, society, or a community in a specific period of time, determined probabilistically as a function of hazard, exposure, vulnerability, and capacity.*

Hazards occur naturally yet become disasters when they interact with society and result in devastation. Exposure refers to locations of people or assets in hazard-prone areas. Vulnerability refers to conditions of people, assets, or systems that increase their susceptibility to hazard impacts. Capacity refers to the knowledge, skills, and resources that individuals and organizations possess and can help reduce disaster risk. Risk is the probability of a future hazard, which cannot be predicted with accuracy. Risk is always present, but its level changes over time and it is unevenly distributed. Factors such as lax building regulations, poverty, and socio-cultural beliefs influence the level of disaster risk. For example, low-income groups from colonized countries are more vulnerable to hazards due to economic poverty, weaker housing, spatial isolation if they live in remote areas where emergency services or internet connectivity is low, limited social capital, and powerlessness (Chambers & Conway, 1992). External factors such as population growth, geopolitical instability, and lack of government investment in risk reduction or early warning systems can compound disaster risks.

This refined comprehension of disaster risk debunks the age-old view of disasters as natural phenomena or divine acts, showing how they are in fact manifestations of intricate and often unpredictable interactions within and between societal systems and environmental triggers. Reflecting shifting perceptions of disaster causation, away from natural constructs and towards a recognition of socionatural roots, current disaster risk reduction for building resilience is oriented towards proactive, systemic strategies rather than reactive, isolated measures.

Concurrently, the concept of "Build Back Better" (BBB) was introduced into the mainstream after the 2004 Indian Ocean Tsunami by the former U.S. President Bill Clinton, to weave disaster risk reduction into recovery efforts (Office of the UN Secretary-General's Special Envoy for Tsunami Recovery, 2006). Its efficacy, however, hinges on nuanced interpretation and implementation, tailored to fit unique societal and environmental contexts. Its novelty and usefulness has been challenged by Maly (2017), Vahanvati and Rafliana (2019), and Cheek and Chmutina (2022), who argued the concept allowed for various interpretations, including rebuilding slightly safer housing or infrastructure, not necessarily informing changes in social and political systems that are the root causes of risk creation. For example, a house rebuilt in a flood zone, with slightly improved structure such as a raised floor level, may still face the risk of damage from future floods. Such incremental betterment or risk-reduction measures are often beyond individual capacities; fundamental changes require policy makers to amend land-use planning flood overlays, integrate water management, and/or support relocation programs. Mere incremental changes are no longer effective; we need fundamental changes within the housing provision systems. Despite its limitations, the concept of BBB has persisted and is increasingly used in practice (Lyons et al., 2010). Transformative (continual and radical) changes, rather than marginal changes, through housing recovery are increasingly urgent, given pressing needs imposed by climate change. UNDRR endorsed BBB, defining it as:

*Using the recovery, rehabilitation, and reconstruction phases after a disaster to increase the resilience of nations and communities through integrating disaster risk reduction measures into the restoration of physical infrastructure and societal systems, and into the revitalization of livelihoods, economies, and the environment.*

As per this definition, BBB goes beyond physical planning, building design, and construction of homes, by promoting multiple recoveries—physical, economic, social, and environmental—alongside risk-reduction measures. Systems thinking intertwined with the "Build Back Better" philosophy has the potential to serve as a catalyst for fundamental changes in housing recovery.

## The Concept of Resilience

The nuanced and multifaceted concept of resilience emerged as a critical element in the discourse on disaster studies, particularly since the 1970s. Recognized under one of the four key priorities within the Sendai Framework for Disaster Risk Reduction (2015–2030), resilience is not merely about "bouncing back" or "rebounding" as per its Latin roots; it entails the capacity of systems, communities, or societies to anticipate, prepare for, prevent, and adapt to changing conditions and to recover from disasters in a way that reduces vulnerability to future events and facilitates sustainable development (UNISDR, 2015). UNDRR (2024) defines resilience as:

*The ability of a system, community, or society exposed to hazards to resist, absorb, accommodate, adapt to, transform, and recover from the effects of a hazard in a timely and efficient manner, including through the preservation and restoration of its essential basic structures and functions through risk management.*

Climate Change studies by IPCC (WGII, 2022, Chapter 1, p. 4) defines resilience as referring to:

*either a process or outcome, encompass not just the concept of maintaining essential function, identity, and structure, but also maintaining a capacity for transformation. Such transformations bring forth questions of justice, power, and politics.*

As noted in these definitions, resilience transcend disciplinary confines, advocating for systems thinking and proactive, long-term risk reduction; it encapsulates the capacities and the intricate interactions between human societies, climate, and biodiversity (IPCC, 2022). The potential of the resilience concept to instigate continuous and transformative change, in coupled socio-ecological systems, is both exciting and challenging, as it confronts issues of justice, power, and politics, inherent in the recovery process.

Housing, from a coupled socio-ecological systems resilience perspective, would incorporate radical changes in hard assets (housing, services, neighborhoods, infrastructure) and soft assets (skills and capacities of individuals, communities, institutions), continually (before, during and after disasters) and at multiple scales.

However, operationalizing resilience, especially within the housing sector, presents unique challenges due to the diversity of interpretations and complexity of embedding such a holistic concept into practice. The term's engineering connotations, referring to the robustness or resistance of structures (Mc Aslan 2010 in IFRC, 2008), may conflict with social science perspectives that focus on community agency, forecasting or anticipatory capacity, and self-organization during disruptions (Imperiale & Vanclay, 2016; Tobin, 1999). Natural sciences' view of resilience as the capacity of ecosystems to retain functionality despite disturbances further adds to the complexity (Holling, 1973). There are at least 28 different definitions of resilience (O'Brien & O'Keefe, 2013), and researchers have noted multiple characteristics (Vahanvati, 2018a). Allen et al. (2019) conceptualized resilience as a process, with a rate (speed) over time, and diverse properties, further adding to the complexity.

Critics of the resilience concept point out its tendency to lack a normative dimension (Bahadur & Tanner, 2014; Davoudi et al., 2012), which can lead to promoting individualistic responses at the expense of community or collective resilience. Additionally, concerns about perpetuating the neoliberal status quo suggest that without a careful and inclusive approach, efforts to build resilience could inadvertently reinforce existing systemic vulnerabilities and social inequalities (Klein, 2008). Despite such complexities, the concept also provides a promising pathway to do things differently and engage in proactive, ex-ante risk-reduction strategies through systemic thinking. A few scholars (Smit & Wandel, 2006) urge us to answer questions about the nature of resilience—of what, to what, when, and for whom—in order to operationalize it.

### *Distinguishing Between Recovery and Resilient Recovery (Resilience of What and Against What?)*

In light of these complexities, we focus on "resilience of" the housing system (housing as a system and housing embedded within a broader system) (Gunderson et al. 2002). We envision "resilience against" all disruptions (anticipated and unexpected, natural hazard, or climate-related). We define resilient recovery (more than housing reconstruction) as:

> A process which increases the capacity of housing systems—including physical structures as well as services, and the capacity of residents inhabiting them—as well as the capacities of the broader system at multiple levels—including communities, governments and institutions, updated policies and standards that govern housing safety and environmental impacts, and a construction sector that ensures labour skills—to undergo systemic change for disaster resilience.

The distinctions between recovery, recovery that is "building back better," and resilient recovery are crucial. Recovery often aims to restore pre-disaster conditions, sometimes perpetuating vulnerabilities. Despite the good intentions behind the idea of "building back better," its lack of nuanced understanding and implementation may result in merely slight improvements to some aspects of pre-disaster conditions. However, resilient recovery seeks to continually learn, transform and improve upon the past to reduce vulnerabilities. This book argues for a paradigm shift towards proactive resilience building, taking lessons from past recoveries to inform more strategic, inclusive, and sustainable future initiatives.

### *Resilience: For Whom?*

Concerning accountability, the question "resilience for whom" is fundamental for those involved in disaster recovery. Reflecting broader developmental changes, the evolution of housing recovery approaches underscores shifts towards more collaborative processes engaging affected communities in a central role, as well as embracing green reconstruction and circular design principles. Encapsulated in international standards such as the *Sphere Handbook* (2018), stakeholders are urged to be held accountable to beneficiaries as the primary stakeholders, and approaches to be humanitarian, not opportunistic. Both the *Sphere Handbook* (2018) and John Turner (1976) in his book *Whose participation in whose decisions and whose actions?* emphasize that the primary responsibility for disaster recovery lies with the state, who ought to encourage collaboration with civil society organizations (CSOs), the private sector, and emergency management authorities, and also balance external aid and the self-recovery capacities of local people. These policies and scholarly discourse are supported by a growing acknowledgment of people being at the center, and using socio-ecological systems approaches for resilient recovery as an imperative of our times.

### *Resilience: When?*

Regarding the timing of resilience across the timeline of recovery, broad consensus holds that it is continuously present, evident as anticipatory resilience pre-disaster, responsive resilience during a disaster, and adaptive resilience long-term post-disaster (Kapucu et al., 2013; Walker & Salt, 2006). Moreover, these stages of resilience building are closely interrelated. For instance, individuals with hazard-safe houses and social networks pre-disaster tend to recover faster after a disaster. Having said that, the crucial post-disaster phase can serve as a "tabula rasa" (Cuff, 2009), literally and metaphorically, providing fertile grounds for new thinking and opening up the possibilities for resilient development pathways.

While the road to achieving resilient recovery is undeniably fraught with challenges, the urgency of climate change and lessons from past recoveries compel an unwavering dedication to systemic change. Several scholars (Bosher, 2010; Folke, 2006; Handmer & Dovers, 1996; Walker & Salt, 2012) underscore the importance of the post-disaster context, in which by adopting proactive measures and centering affected communities, housing can lay the foundation for a more resilient future, fundamentally altering the trajectory of societal development and hazard adaptability.

### *Resilience Characteristics*

Revisiting the attributes critical to resilience offers a clearer understanding of what makes housing not just a shelter, but a stronghold against the impacts of hazards through systemic changes. The term "resilience" encompasses various characteristics (Table 1.1); when related to housing and its inhabitants, these form a multifaceted portrait of safety, adaptability, and inclusivity.

**Robustness (Engineering Resilience):** The robustness of a housing system is its capacity to withstand stress or shock (Bevc, 2013; Folke, 2006; IFRC, 2012). Robustness is reflected in the broader context of land-use planning and building regulations that dictate exposure of housing, structural design, and construction quality, ensuring that houses can endure the predicted intensity of potential hazards (Davis & Alexander, 2016; Schilderman & Parker, 2014). Robustness is the bedrock principle that safeguards dwellings and their inhabitants from imminent threats.

**Appropriateness and Adaptability:** A resilient home goes beyond structural integrity—its social value, aesthetic, and cultural fit make it truly appropriate. Comfort, especially thermal comfort, and the flexibility of a house for future modifications enhance its adaptability. Guided by principles of sustainability, developed by UNEP and SKAT (2007), housing recovery must also pivot towards reducing construction waste and embracing reusable and recyclable components. Appropriate and adaptable homes are welcoming spaces that can be adapted, extended, or evolve over time, with the changing needs of their residents.

**Integration and Redundancy (buffer capacities):** A dwelling functions not in isolation but within an integrated system with interconnected subsystems of water, electricity, sewage, and drainage. This integration extends to the broader infrastructure network of roads, street lighting, and educational and recreational facilities (Charlesworth & Ahmed, 2015; Vahanvati, 2018a). Redundancy, or buffer capacity, involves deliberately designing systems with a safe margin for failure. This "spare capacity" (O'Connell et al., 2020) is envisaged to accommodate unforeseen pressures without a cascade of disruption. This "safe-to-fail" approach (Davis & Alexander, 2016) is a deliberate redundancy to ensure continuity of daily functioning, in the face of adversity. Examples include stockpiling emergency resources, backup power suppliers (IFRC, 2012), or multiple

**Table 1.1** Key characteristics for assessing resilience outcomes in the long-term after post-disaster housing reconstruction projects (Source: Authors)

*Intended outcome: enhanced capacities in housing as a system, within a system*

| *Resilience characteristics* | *Resilience outcomes in housing systems over the long term* | *Description of long-term resilience outcomes in housing, communities, and institutions* |
|---|---|---|
| Robustness and diversity (buffer capacities) | Planning and building standards | – Well located for reduced exposure<br>– Building standards stronger/revised |
| | Structural safety | – Structural hazard-safety features explicitly integrated in housing<br>– Houses withstood another disruption<br>– Good construction quality<br>– Choice of materials and construction technologies<br>– Traditional/local construction labor incorporated<br>– Replication of improved construction systems |
| Appropriateness and adaptability | House design | – Design with diverse housing typologies<br>– Culturally responsive spatial layout<br>– Thermally comfortable design<br>– Design flexibility for extension or adaptation<br>– Occupancy in houses |
| Redundancy and integration (buffer capacities) | Settlement or city development | – Services integrated with housing<br>– Infrastructure integrated with settlement<br>– Social amenities integrated in settlement<br>– Livelihood proximity/considerations<br>– Redundancy integrated in housing or settlement |
| Rights, equity, and security | Equity, inclusion, dignity, affordability, and security | – Recovery for all residents<br>– Participatory process<br>– Low cost, affordable and ease of maintenance<br>– Better security of land, housing tenure |
| Resourcefulness (adaptive capacities) | Institutional capacities: recovery program, assistance, agility, rapidity, timely yet strategic | – Enforcement of planning and building regulations<br>– Improved governance and institutional capacity<br>– Proactive investment in prevention strategies |
| | Social capacities<br>Individual capacities | – Changed social mindset, norm, and improved or preserved social bond<br>– Improved household awareness, security, and reduced aid dependency<br>– Better quality of life and political voice |

layers of coastal defenses such as mangroves, sea walls, or levees built to minimize disruption from storm surges, even if one layer fails.

**Rights and Equity:** Equity in housing implies all-encompassing inclusivity, where the rights and needs of vulnerable groups, including children, women, elderly, people with disabilities, and the economically poor (Mulligan et al., 2012), are prioritized. Global initiatives like disability-inclusive DRR (DiDRR), gender-sensitive DRR (UNISDR, 2015), and inclusive DRR for resilience building (ADPC, 2018), emphasize that everyone must have equitable access to safe housing. Community-driven recovery interventions harness the potential for profound social change, bridging divides and promoting a synthesis of diverse perspectives for more robust recovery outcomes (Mulligan et al., 2012; Oliver-Smith, 2010).

**Resourcefulness (adaptive capacities):** Housing resilience is also characterized by resourcefulness—the aggregate of peoples' or organizations' abilities, strengths, and access to resources (social, financial, or technical) that enable communities to manage and reduce risks (UNDRR, 2024). Abilities and capacities are foundational for building resilience. Adaptive capacities are defined as "the ability of humans (individuals, collectives, or institutions) to imagine the future, [and] the capacity for forward planning" (Holling & Walker, 2003, p. 2). Institutional capacities include recovery programming, resource mobilization and coordination, and organizational setup (Bevc, 2013; Davis & Alexander, 2016; Smit & Wandel, 2006; Twigg, 2009). Resourcefulness thus embodies the collective capacity to envisage the future and strategize adaptively. Resourcefulness in the long term after housing recovery translates into integrating vital lessons into risk-aware policies, thus reinforcing preparedness and bolstering the adaptive capacity of housing systems (Fig. 1.1).

## The Essence of Resilient Recovery: An Imperative of Our Times

Resilient recovery transcends the act of reconstruction; it signifies a holistic transformation of the housing system. This process integrates both tangible (hard) assets—such as buildings, structures and infrastructure—and intangible (soft) assets—like the knowledge, skills, capacities, resources, abilities, values, and cultural practices of the residents—ensuring a synergistic evolution of the entire housing system. Resilience in housing encompasses revising land-use plans, building regulations, and strengthening institutional

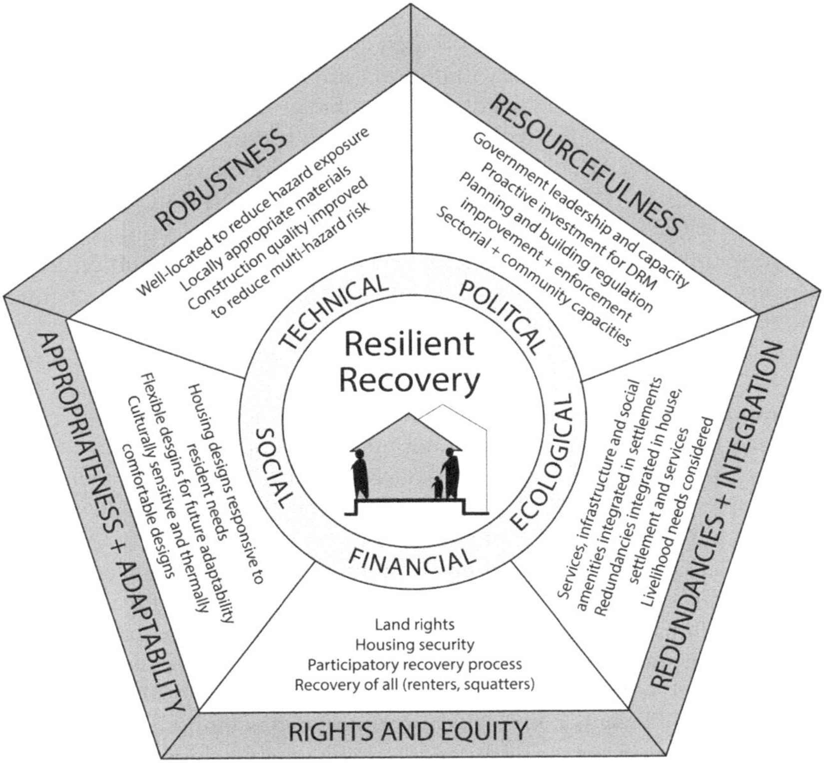


**Fig. 1.1** Resilience characteristics conceptual framework. (Source: M. Vahanvati with feedback from the authors)

capacities to address and reduce vulnerabilities while simultaneously augmenting communities' and individuals' strengths, capacities, and capabilities in making risk-informed decisions for confronting future hazards.

Drawing lessons from past recoveries informs wiser, more proactive future strategies that consciously avoid past missteps. Housing recovery has matured from a passive, reactive approach towards fostering proactive resilience, echoing broader developmental transformations. These paradigm shifts—from autocratic to participatory, from preservation to progress—mirror changes in international policy, specifically those spearheaded by the United Nations' (UNDRR, 2021; UNISDR, 1994, 2005, 2015) pursuit of disaster risk reduction. These shifts in strategy, transitioning from a singular focus on structural integrity and economic recovery towards comprehensive human-centered socio-ecological system resilience, have contributed to the broader adoption of the resilience concept.

Navigating the journey from mere reconstruction to resilient recovery is fraught with challenges of power and politics; however, the urgency brought on by climate change compels a steadfast dedication to overcoming these hurdles. Housing serves as a cornerstone upon which individual and national resilience can be built. Resource constraints, governance issues, and entrenched socio-cultural mindsets pose significant obstacles. Overemphasis on technical solutions at the expense of cultural relevance, or reliance on financial aid without strategic oversight, have proven inadequate.

The theory of change underpinning resilient recovery advocates for multiple layers of transformation. For individuals, it means cultivating a deeper understanding of risk and risk-informed decision making about safe housing practices. Communities undergo a cultural reimagining, increasingly valuing indigenous architecture and construction methods that harmonize with local environments. Civil society organizations and private sector participants in disaster recovery pivot towards collaborative, culturally sensitive, and climate-adaptive designs. The construction industry adopts practices that lower energy and material usage, favoring local, durable, and low-impact resources while capitalizing on indigenous skills and minimizing waste.

Institutional change is paramount. It involves a commitment to continuous learning, capturing lessons from past recoveries and application of best practices in disaster risk reduction. Proactively updating building codes and standards and revising land-use plans to reflect new hazard risk data are essential steps in ensuring future housing resilience (Comerio, 1998, 2014). Moreover, it is incumbent upon governments worldwide to champion updates for building codes that embrace indigenous technologies and green standards (Cuff, 2009; Lisa, 2010; Orr, 2004; Papanek, 1995), underpinning a zero-carbon, green, affordable, and culturally attuned future of housing.

Recognizing three archetypes of recovery—restorative recovery, progressive recovery (build back better), and transformative recovery (resilient recovery)—clarifies their distinct impacts. Restorative recovery simply returns to the status quo, while progressive recovery seeks minor enhancements. Transformative recovery, however, is committed to systemic change, earnestly reducing the risk from future hazards and shifting society's development towards a resilient trajectory.

## References

ADPC. (2018). *Building Resilience through Inclusive and Climate-Adaptive Disaster Risk Reduction in Asia-Pacific (BRDR Program)*. Asian Disaster Preparedness Centre. Retrieved from http://www.adpc.net/igo/contents/publications/publications-Details.asp?pid=1353#sthash.xC8vTwhI.WrhBGZ65.dpbs

Allen, C. R., Angeler, D. G., Chaffin, B. C., Twidwell, D., & Garmestani, A. (2019). Resilience reconciled. *Nature Sustainability 2*(10), 898–900. https://doi.org/10.1038/s41893-019-0401-4

Bahadur, A., & Tanner, T. (2014). Transformational Resilience Thinking: Putting People, Power and Politics at the Heart of Urban Climate Resilience. *Environment & Urbanization, 26*(1), 200–214. https://doi.org/10.1177/0956247814522154

Barakat, S. (2003, December). Housing Reconstruction After Conflict and Disaster. *Humanitarian Practice Network, 43*. Retrieved August 25, 2018, from http://www.odihpn.org/documents/networkpaper043.pdf

Barenstein, J. D., & Iyengar, S. (2010). India: From a Culture of Housing to a Philosophy of Reconstruction. In M. Lyons, T. Schilderman, & C. Boano (Eds.), *Building Back Better: Delivering People-Centred Housing Reconstruction at Scale* (pp. 163–190). Practical Action. Retrieved from https://practicalaction.org/print/disaster-reduction/sudan/access-to-services/docs/ia3/building-back-better-lyons-schilderman.pdf#page=175

Barenstein, J. E., & Leemann, E. (2012). *Post-Disaster Reconstruction and Change: Communities' Perspectives*. CRC Press.

Bevc, C. (2013). Introduction to Conceptual Insights and Applications of Resilience. In N. Kapucu, C. V. Hawkins, & F. I. Rivera (Eds.), *Disaster Resiliency* (pp. 15–20). Routledge.

Blasius, W. (1877). Causes of the Huron Disaster. *Proceedings of the American Philosophical Society, 17*(100), 212–215.

Bosher, L. (2010). The Importance of Institutional and Community Resilience in Post-disaster Reconstruction. In G. Lizarralde, C. Johnson, C. Davidson, & Ebooks Corporation (Eds.), *Rebuilding after Disasters from Emergency to Sustainability* (pp. 231–247). Spon Press.

Cambridge Dictionary. (2023). Development. In *Meaning of Development*. Retrieved May 18, 2022, from https://dictionary.cambridge.org/dictionary/english/development

Carson, R. (2002). *Silent Spring* (40th anniversary ed., 1st Mariner Books ed. ed.). Houghton Mifflin.

Chambers, R. (1983). *Rural Development: Putting the Last First*. Taylor and Francis.

Chambers, R. (1995). Poverty and Livelihoods: Whose Reality Counts? *Environment and Urbanization, 7*(1), 173–204.

Chambers, R., & Conway, G. R. (1992). Sustainable Rural Livelihoods: Practical Concepts for the 21st Century. In *Institute of Development Studies Discussion Paper 296*. Retrieved January 1, 1992, from RePEc.

Charlesworth, E., & Ahmed, I. (2015). *Sustainable Housing Reconstruction: Designing Resilient Housing after Natural Disasters* (1st ed.). London, Routledge. https://doi.org/10.4324/9781315735412

Cheek, W., & Chmutina, K. (2022). 'Building Back Better' is Neoliberal Post-disaster Reconstruction. *Disasters, 46*(3), 589–609. https://doi.org/10.1111/disa.12502

Comerio, M. C. (1998). *Disaster Hits Home: New Policy for Urban Housing Recovery*. University of California Press. Retrieved from http://www.h-net.org/review/hrev-a0b1z3-aa

Comerio, M. C. (2014). Disaster Recovery and Community Renewal: Housing Approaches. *Cityscape (Washington, D.C.), 16*(2), 51–68.

CRED EM-DAT. (2009). *Natural Disasters Trends* [Graph]. CRED, Emergency Events Database (EM-DAT). Retrieved from www.emdat.be/natural-disasters-trends

Cuff, D. (2009). Design after Disaster [Feature Article]. *Places: Forum of Design for the Public Realm, 21*(1), 4–7. Retrieved from http://search.ebscohost.com/login.aspx?direct=true&db=aft&AN=505408915&site=ehost-live&scope=site

Davis, I. (1978). *Shelter After Disaster*. Oxford Polytechnic Press.

Davis, I., & Alexander, D. (2016). *Recovery from Disaster*. Taylor and Francis. https://doi.org/10.4324/9781315679808

Davoudi, S., Shaw, K., Haider, L. J., Quinlan, A. E., Peterson, G. D., Wilkinson, C., Fünfgeld, H., McEvoy, D., & Porter, L. (2012). Resilience: A Bridging Concept or a Dead End? "Reframing" Resilience: Challenges for Planning Theory and Practice Interacting Traps: Resilience Assessment of a Pasture Management System in Northern Afghanistan Urban Resilience: What Does it Mean in Planning Practice? Resilience as a Useful Concept for Climate Change Adaptation? The Politics of Resilience for Planning: A Cautionary Note: Edited by Simin Davoudi and Libby Porter. *Planning Theory & Practice, 13*(2), 299–333. https://doi.org/10.1080/14649357.2012.677124

Folke, C. (2006). Resilience: The Emergence of a Perspective for Social–Ecological Systems Analyses. *Global Environmental Change, 16*(3), 253–267. https://doi.org/10.1016/j.gloenvcha.2006.04.002

Grant, W. (2019). *How We Discovered the Climate Problem*. Australian National University. Retrieved August 19, from https://www.policyforum.net/how-we-discovered-the-climate-problem/

Gunderson, L., Folke, C., Lee, M., & Holling, C. S. (2002). Memory of Mavericks. *Conservation Ecology, 6*(2), N/a. https://doi.org/10.5751/ES-00423-060219

Handmer, J. W., & Dovers, S. R. (1996). A Typology of Resilience: Rethinking Institutions for Sustainable Development. *Industrial & Environmental Crisis Quarterly, 9*(4), 482–511. https://doi.org/10.1177/108602669600900403

Holling, C. (1973). Resilience and Stability of Ecological Systems. *Annual Review of Ecology and Systematics, 41*(1), 1–23. https://doi.org/10.1146/annurev.es.04.110173.000245

Holling, C., & Walker, B. (2003). Resilience Defined. In C. S. Holling & B. Walker (Eds.), *Internet Encyclopedia of Ecological Economics*. International Society for Ecological Economics.

Hunnarshala. (2007). *Three Themes—Community Driven Housing*. Hunnarshala Foundation, India. Retrieved from www.hunnarshala.org/about-us.html

IFRC. (2008). *A Framework for Community Safety and Resilience*. International Federation of Red Cross and Red Crescent Societies. International Federation of Red Cross and Red Crescent Societies (IFRC). Retrieved from www.ifrc.org/Global/Case%20studies/Disasters/cs-framework-community-en.pdf

IFRC. (2010). *Owner Driven Housing Reconstruction Guidelines*. International Federation of Red Cross & Red Crescent Societies (IFRC). Retrieved September, from www.ifrc.org/PageFiles/95526/publications/E.02.06.%20ODHR%20Guidelines.pdf

IFRC. (2012). *Understanding Community Resilience and Program Factors that Strengthen Them: A Comprehensive Study of Red Cross and Red Crescent Societies Tsunami Operation*. International Federation of Red Cross and Red Crescent Societies. IFRC. Retrieved from https://reliefweb.int/report/indonesia/understanding-community-resilience-and-program-factors-strengthen-them

Imperiale, A. J., & Vanclay, F. (2016). Experiencing Local Community Resilience in Action: Learning from Post-disaster Communities. *Journal of Rural Studies, 47*, 204–219. https://doi.org/10.1016/j.jrurstud.2016.08.002

IPCC. (2022). *AR6 WGII Climate Change Impacts, Adaptation and Vulnerability*. IPCC. Retrieved from https://www.ipcc.ch/report/ar6/wg2/

IPCC. (2023). *Climate Change 2023: Synthesis Report*. IPCC. Retrieved from https://www.ipcc.ch/report/ar6/syr/downloads/report/IPCC_AR6_SYR_LongerReport.pdf

Jha, A. K., Barenstein, J. D., Phelps, P. M., Pittet, D., & Sena, S. (2010). *Safer Homes, Stronger Communities: A Handbook for Reconstructing after Natural Disasters*. The World Bank. Retrieved from www.housingreconstruction.org/housing/sites/housingreconstruction.org/files/Disaster%20Types%20and%20Impacts.pdf

Kapucu, N., Hawkins, C. V., & Rivera, F. I. (2013). *Disaster Resiliency: Interdisciplinary Perspectives*. Routledge.

Klein, N. (2008). *The Shock Doctrine: The Rise of Disaster Capitalism* ([Penguin ed.]. ed.). Penguin.

Klein, N. (2014). *This Changes Everything: Capitalism vs. the Climate.* Simon & Schuster.

Lisa, F. (2010). *Green Rehabilitation: Environmental Sustainability and Post-disaster Shelter in India* [Dissertation, Oxford Brookes University]. Unpublished. Retrieved from http://oisd.brookes.ac.uk/architecture/cendep/dissertations/FedericaLisa.pdf

Lizarralde, G., Johnson, C., Davidson, C. H., & Ebooks Corporation. (2010). *Rebuilding after Disasters from Emergency to Sustainability.* Spon Press. Retrieved from http://www.RMIT.eblib.com.au/EBLWeb/patron?target=patron&extendedid=P_446574_0&

Lyons, M., Schilderman, T., & Boano, C. (2010). *Building Back Better: Delivering People-Centred Housing Reconstruction at Scale.* Practical Action.

Maly, E. (2017). Rethinking "build Back Better" in housing reconstruction: A proposal for "people Centered Housing Recovery". IOP Conference Series. *Earth and Environmental Science 56*(1), 12025. https://doi.org/10.1088/1755-1315/56/1/012025.

Mulligan, M., Ahmed, I., Mercer, D., Nadarajah, Y., & Shaw, J. (2012). Long Term Social Recovery Following the 2004 Tsunami: Community, Livelihoods, Tourism and Housing. *Environmental Hazards: Human and Policy Dimensions, 11*(1), 38–51.

O'Brien, G., & O'Keefe, P. (2013). *Managing Adaptation to Climate Risk: Beyond Fragmented Responses.* Routledge. https://doi.org/10.4324/9780203836910

O'Connell, D., Grigg, N., Maru, Y., Bohensky, E., Hayman, D., Measham, T., Wise, R., Dunlop, M., Patterson, S., Vaidya, S., & Lade, S. (2020). *A Resilience Checklist—A Guide for Doing Things Differently and Acting Collectively.* CSIRO. Retrieved from https://www.qra.qld.gov.au/sites/default/files/2020-11/a_resilience_checklist_-_a_guide_for_doing_things_differently_and_acting_collectively.pdf

Office of the UN Secretary-General's Special Envoy for Tsunami Recovery. (2006). *Lessons Learned from Tsunami Recovery—Key Propositions for Building Back Better.* Retrieved from www.preventionweb.net/files/2054_VL108301.pdf

OHCHR, & UN-Habitat. (2009). *The Right to Adequate Housing.* Retrieved from www.ohchr.org/EN/Issues/Housing/toolkit/Pages/RighttoAdequateHousingToolkit.aspx

Oliver-Smith, A. (1996). Anthropological Research on Hazards and Disasters. *Annual Review of Anthropology, 25*(1), 303–328. https://doi.org/10.1146/annurev.anthro.25.1.303

Oliver-Smith, A. (2010). *Defying Displacement Grassroots Resistance and the Critique of Development* (1st ed.). University of Texas Press.

Orr, D. (2004). *Earth in Mind: On Education, Environment, and the Human Prospect*. Island Press.

Papanek, V. (1995). *The Green Imperative: Ecology and Ethics in Design and Architecture*. Thames and Hudson.

Roser, M. (2021). *The Economies that are Home to the Poorest Billions of People Need to Grow if we Want Global Poverty to Decline Substantially*. Our World in Data. Retrieved from https://ourworldindata.org/poverty-growth-needed#article-citation

Sachs, W. (2010). *The Development Dictionary: A Guide to Knowledge as Power* (2nd ed.). Zed Books.

Schilderman, T., & Parker, E. (2014). *Still Standing? Looking Back at Reconstruction and Disaster Risk Reduction in Housing*. Practical Action.

Schumacher, E. F. (1993). *Small is Beautiful: A Study of Economics as if People Mattered*. Vintage.

Sen, A. (1985). *Commodities and Capabilities*. North-Holland Elsevier Science.

Smit, B., & Wandel, J. (2006). Adaptation, Adaptive Capacity and Vulnerability. *Global Environmental Change, 16*(3), 282.

Sphere Handbook. (2018). *Humanitarian Charter and Minimum Standards in Humanitarian Response* (4th ed.). Sphere Association. Retrieved from www.spherestandards.org/handbook

Tobin, G. A. (1999). Sustainability and Community Resilience: The Holy Grail of Hazards Planning? *Global Environmental Change B: Environmental Hazards, 1*(1), 13–25. https://doi.org/10.1016/S1464-2867(99)00002-9

Turner, J. F. (1976). *Housing by People: Towards Autonomy in Building Environments*. Marion Boyars.

Twigg, J. (2009). *Characteristics of a Disaster Resilient Community* [Guidance Note]. Retrieved October 18, 2012, from http://practicalaction.org/docs/ia1/characteristics-disaster-resilient-community-v2.pdf

UNDP. (2022). *Handbook on Owner-Driven Housing Reconstruction*. Retrieved from https://www.preventionweb.net/publication/handbook-owner-driven-housing-reconstruction

UNDRO. (1982, 2010). *Shelter After Disaster: Guidelines for Assistance*. UNDRO. Retrieved from http://humanitarianlibrary.org/sites/default/files/2014/02/Shelter_After_Disaster_UNDRO.pdf

UNDRR. (2021). *UNDRR Strategic Framework 2022–2025*. UN Office for Disaster Risk Reduction.

UNDRR. (2022). *Global Assessment Report on Disaster Risk Reduction: Our World at Risk: Transforming Governance for a Resilient Future*. UNDRR. Retrieved from https://www.undrr.org/gar2022-our-world-risk

UNDRR. (2024). *Terminology on Disaster Risk Reduction*. UNISDR (United Nations International Strategy for Disaster Reduction). Retrieved from https://www.undrr.org/terminology

UNEP, & SKAT. (2007). *After the Tsunami: Sustainable Building Guidelines for South-East Asia*. U. a. SKAT. Retrieved from www.ifrc.org/PageFiles/95891/E.02.05.After%20Tsunami.%20Building%20Guidelines%20SE%20Asia%20UNEP-SKAT.pdf

UN-Habitat. (1976). *Habitat I*. United Nations. Retrieved March 20, 2021, from http://habitat1.org/ and https://www.un.org/en/conferences/habitat/vancouver1976

UN-Habitat. (2003). *The Challenge of Slums: Global Report on Human Settlements*. United Nations Human Settlements Programme. Retrieved from https://unhabitat.org/sites/default/files/download-manager-files/The%20Challenge%20of%20Slums%20-%20Global%20Report%20on%20Human%20Settlements%202003.pdf

UN-Habitat. (2022). *Housing*. United Nations Human Settlements Programme. Retrieved November 29, 2022, from https://unhabitat.org/topic/housing

UN-Habitat. (2023). *Housing Matters*. United Nations Human Settlements Programme,. Retrieved November 29, 2023, from https://www.worldurban-campaign.org/housingmatters

UNISDR. (1994). *Yokohama Strategy and Plan of Action for a Safer World*. UNISDR (United Nations Office for Disaster Risk Reduction). Retrieved from www.unisdr.org/we/inform/publications/8241

UNISDR. (2005). *Hyogo Framework for Action 2005–2015: Building the Resilience of Nations and Communities to Disasters*. UNISDR (United Nations Office for Disaster Risk Reduction). Retrieved from www.preventionweb.net/files/594_10382.pdf

UNISDR. (2015). *Sendai Framework for Disaster Risk Reduction 2015–2030*. UNISDR (United Nations Office for Disaster Risk Reduction). Retrieved from www.unisdr.org/we/coordinate/sendai-framework

United Nations General Assembly. (2017). *New Urban Agenda* (HABITAT III: The United Nations Conference on Housing and Sustainable Urban Development, Issue). Retrieved from https://habitat3.org/the-new-urban-agenda/

Vahanvati, M. (2018a). A Novel Framework for Owner Driven Reconstruction Projects to Enhance Disaster Resilience in the Long Term. *Disaster Prevention and Management, 27*(4), 421–446. https://doi.org/10.1108/DPM-11-2017-0285

Vahanvati, M. (2018b). *Thesis: Owner-Driven Housing Reconstruction as a Means of Enhancing Disaster Resilience of At-risk Communities in India*. RMIT University. Retrieved from https://researchrepository.rmit.edu.au/discovery/fulldisplay/alma9921864052701341/61RMIT_INST:ResearchRepository?tags=scholar

Vahanvati, M., & Rafliana, I. (2019). Reliability of Build Back Better at Enhancing Resilience of Communities. *International Journal of Disaster Resilience in the Built Environment, 10*(4), 208–221. https://doi.org/10.1108/IJDRBE-05-2019-0025

Walker, B., & Salt, D. (2006). *Resilience Thinking: Sustaining Ecosystems and People in a Changing World*. Island Press.

Walker, B. H., & Salt, D. (2012). *Resilience Practice: Building Capacity to Absorb Disturbance and Maintain Function*. Island Press.

Warf, B. (2001). Space and Social Theory in Geography. In N. J. Smelser & P. B. Baltes (Eds.), *International Encyclopedia of the Social & Behavioral Sciences* (pp. 14743–14749). Pergamon. https://doi.org/10.1016/B0-08-043076-7/02498-0

WGII, I. A. (2022). *AR6 WGII Chapter 1 Climate Change Impacts, Adaptation and Vulnerability*. IPCC. Retrieved from https://www.ipcc.ch/report/ar6/wg2/

Winchester, P. (1979). Disaster Relief Operations In Andhra Pradesh, Southern India, Following The Cyclone In November 1977. *Disasters, 3*(2), 173–177. https://doi.org/10.1111/j.1467-7717.1979.tb00222.x

Wisner, B., Blaikie, P., Cannon, T., & Davis, I. (2004). *At Risk: Natural Hazards, People's Vulnerability and Disasters* (2nd ed.). Routledge. Retrieved from www.preventionweb.net/files/670_72351.pdf

CHAPTER 2

# Long Term Post-Disaster Housing Reconstruction in India: Case Studies (Lower-Middle-Income Country)

**Abstract** In this chapter, the authors aim to identify key factors that helped or hindered systemic changes through post-disaster housing reconstruction projects, from a rights and resilience perspective over the long term (at least 10 years since the disasters) in India. Following an explanation of India's disaster management and housing context, we explain two housing reconstruction projects as case studies after each disaster in India—the 2001 Gujarat Earthquake and the 2008 Kosi River floods in Bihar (see Fig. 2.1). Looking back at these two disasters, which occurred 24 and 17 years ago, respectively, we examine the long-term outcomes of two post-disaster housing reconstruction projects across four critical time periods, including: (i) the pre-disaster housing context, (ii) disaster impacts and recovery policies, (iii) post-disaster implementation and (iv) long-term outcomes. For longitudinal comparative case study analysis, site visits and 80 semi-structured interviews were conducted in 2014–2015 (drawing from past research), and 35 semi-structured interviews in 2022–2023 17 in Gujarat and 18 in Bihar). To attest the validity of data, interviews were conducted with diverse recovery stakeholders including government authorities, grassroots civil society organizations (CSO), practitioners (planners, architects, builders), academics and residents who survived the disaster (beneficiaries and non-beneficiaries). Findings reveal some commonalities and changes in recovery strategies over time, as well as the adoption of lessons from disasters in 2001 to 2008 in India. The conclusions draw on the strengths and weaknesses of India's owner-driven



M. Vahanvati et al., *Resilient Recovery from Disasters*,
https://doi.org/10.1007/978-981-97-8047-1_2

housing reconstruction (ODHR) policies and practices in fostering a more resilient future, offering recommendations for improvements and potential lessons for other countries in relation to negotiations between government and CSOs and integration of technical resilience with socio-ecological resiliency. This chapter reveals India as a nation at a crossroads, where despite a best-practice policy environment, policy translation into practice—the urgency of implementing effective disaster management and resilient housing practices—meets the challenges of continued education, construction skills trainings, enforcement of building standards and empowerment for making risk-informed decisions.

**Keywords** India • Owner-driven • People-centered • Post disaster • Housing reconstruction • Disaster resilience • Long-term recovery • Civil society organizations • 2001 Gujarat Earthquake • 2008 Bihar Kosi River floods

## India's Housing, Hazard, and Disaster Management Context

India's disaster vulnerability is closely linked to risk exposure, diverse geo-climatic conditions as well as socioeconomic vulnerabilities. India is the second-most disaster-prone nation globally, with a World Risk Index (WRI) of 42.31 (Mucke et al., 2022). The nation has a diverse topography with risks from various hazards, including earthquakes, floods, landslides, and cyclones. Earthquakes pose a significant risk, with about 60% of India's landmass located in seismic zones III–IV. Floods affect over 40 million hectares of land (GoI & UNDP, 2011) and landslides affect 12% of India's land. Additionally, cyclones and storm surges impact about 75% of India's coastline. The Himalayan region and Andaman Nicobar Islands are particularly vulnerable to earthquakes and landslides. Some high-profile disasters India has witnessed include the 1999 Orissa Super Cyclone, the 2001 Gujarat Earthquake, the 2004 Indian Ocean Tsunami, and the 2018 Kerala floods. These disasters have cost the country an average of 2% of its GDP annually (GoI & UNDP, 2011), with floods alone costing 0.5% of the GDP (NDMA et al., 2020; WMO, 2021), and have led to displacement and internal migration, compounding challenges for already vulnerable populations.

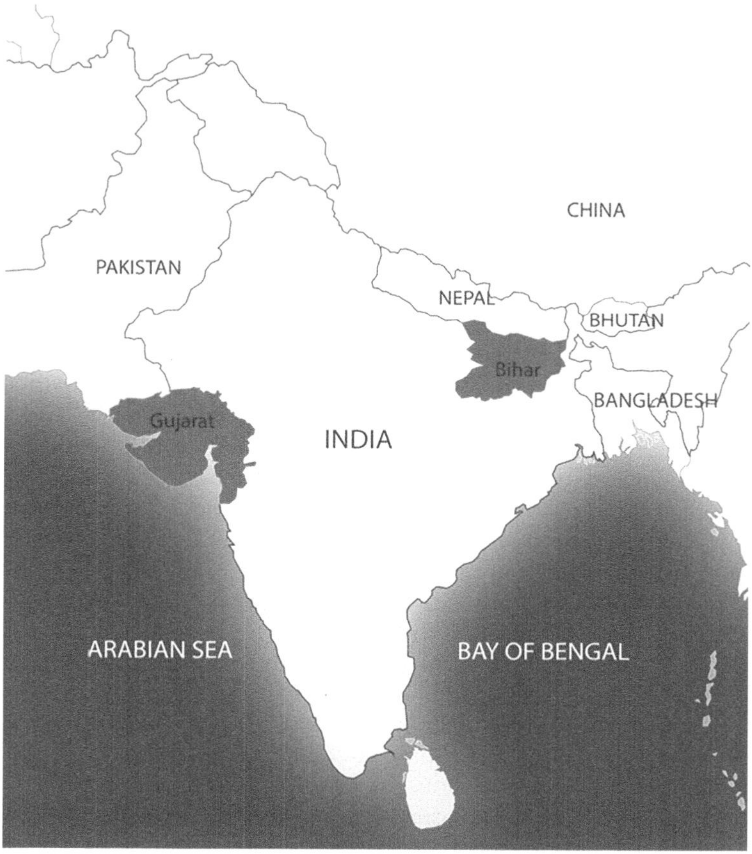


**Fig. 2.1** Map of Indian case studies in the states of Gujarat and Bihar (Source: Authors)

India's socioeconomic vulnerabilities to disasters include extreme economic inequalities or poverty along with deep-rooted social disadvantages due to the caste system, religion, and gender inequality. Despite being one of the world's largest democracies, India's caste system has left a large proportion of the population at a socioeconomic disadvantage. Consequently, the country ranked 132 on the Human Development Index (HDI), in the medium human development category in 2020, dropping from the HDI value from 2019 (UNDP, 2023). The lack of

India's HDI progress reflects development realities on the ground, including high population growth, with India becoming the most populous country in the world as of April 2023 (UN-DESA, 2023). Despite India's rapid economic growth, while transitioning to a post-COVID-19 world a further divide has grown between the rich and the poor. Data from 1922–2023 by the World Inequality Lab highlights that 10% of India's population holds over 60% of its wealth, with the top 1% income share among the highest in the world (Kumar Bharti et al., 2024), producing 70 new millionaires every day, while the bottom 50% have seen a decline in wealth and nearly 20% of the population living in poverty (Suri, 2023). Socioeconomic disadvantages compel people to live in unsafe environments and build sub-standard housing, further magnifying disaster risk.

Climate change is exacerbating these vulnerabilities. India experiences climate-related hazards like heat, droughts, hailstorms, sea level rise, and glacier retreat. India currently stands as the world's third highest greenhouse gas emitter, giving the country a ranking of 138 on the climate vulnerability ND-GAIN Index (UNDP, 2024). Though India committed to a net-zero greenhouse gas emissions target by 2070 at the United Nations Climate Conference in 2021, in Glasgow (COP26), it was shown that the country has a long way to go to address complex geographical exposure and socioeconomic vulnerabilities.

India's housing and infrastructure are strained by rapid urbanization, rapid population growth, land shortages, skills shortages, and socioeconomic inequalities. Limited land availability has led to higher population density, approximately 473 people per square kilometer in 2024 (World Bank, 2024). Moreover, to house a rapidly growing population, India requires a doubling of the housing stock by 2050. The national government has implemented policy initiatives such as the *Pradhan Mantri Awas Yogna (PMAY)*, translated as the Prime Minister's Housing Policy, which provides financial assistance for housing to the poor. The predecessor Housing Policy—*Indira Away Yogna* (1995–2014)—allocated 30 million rupees for rural houses, while PMAY (since 2015) allocated 12 million rupees for urban and 30 million rupees for rural households (interviewee). Despite such efforts, nearly 49% of India's population continues to live in slums (World Bank, 2022). Additionally, India grapples with challenges ensuring construction quality and adhering to building construction standards, despite progressive building regulations and guidelines for engineered (e.g. concrete, steel) as well as non-engineered (e.g. bamboo, mud) construction systems developed by the Bureau of Indian Standards

(Arya et al., 2014). Building codes are adhered to only when professional architects and engineers are involved, equating to the only 30–40% of all building stock built by formal building practitioners (Shah, 2009). Nearly 60% of urban housing stock and the entirety of rural housing stock is built by people themselves or with the help of non-formal builders. Additionally, a lack of formal construction skills training for builders, suppliers, and laborers, who typically learn construction skills informally during construction, and limited enforcement or oversight of compliance with building standards, has led to the sub-standard quality of most of the housing stock in India, which remains at risk of destruction in the face of disasters. Thus, housing in India is a stark reminder of the significance of translating policy into practice for disaster resilience.

## India's Federated Disaster Risk Management and Housing Context

Early Acts, policies, laws, and regulations implemented by the Government of India include a National Centre for Disaster Management, established in 1995, which was renamed the National Institute of Disaster Management (NIDM) in 2003 and became a statutory organization in 2005; and a High Powered Committee established in 1999 (until 2001), against the backdrop of the International Decade for National Disaster Reduction (IDNDR) and the 1999 Odisha Super Cyclone. A turning point was the 2001 Gujarat Earthquake, which led to the creation of the National Disaster Management Act (NDMA, 2005), state- and district-level disaster management authorities (i.e. SDMA, DDMA) and the National Disaster Response Force (NDRF). The 2004 Indian Ocean Tsunami led to further development and refinement through a National Disaster Management Policy in 2009, a Disaster Management India Report in 2011, two National Disaster Management Plans in 2016 and 2019, and the Prime Minister's Ten-Point Agenda in 2016. It has taken over 20 years for India to develop a legal structure with multi-level institutional mechanisms for systematic and holistic disaster management. Due to India being a federation of 28 states and 8 union territories, the Act has created structured institutional mechanisms with NDMA as the apex body, headed by the prime minister at the national level, and corresponding disaster management authorities at state-(SDMA), district-(DDMA), and sub-divisional or block-levels, for facilitating coordinated response and recovery efforts. Additionally, the Government of India also established funding procedures, to be managed through two agencies, as the State Disaster Response Fund (SDRF) and the National Disaster Response Fund (NDRF). Since 2001, India's disaster management approach has evolved from a reactive one (or crisis management) to be a more proactive one by focusing on prevention and mitigation joined by financial mechanisms to ensure that recovery efforts also lead to preventing future hazards from becoming devastating. Despite a proactive disaster management framework, active government partnerships with civil society organizations (CSOs) and the private sector, and a multi-tiered, structured institutional mechanism (joining policies, plans, finances and players), socioeconomic vulnerability reduction, environmental protection, and hazard mitigation continue to remain challenging.

## Case 1: Gujarat Housing Reconstruction Post-2001 Earthquake

The state of Gujarat, located on the western edge of India and sharing a border with Pakistan, is known as the textile hub of India and the place where Mohandas (Mahatma) Gandhi was born. Geo-climatically, the state has varied terrains, ranging from the salt deserts of Kachchh with generally arid and semi-arid scrublands of the western parts, to fertile, coastal plains of the southeastern parts. The region has mainly two forms of livelihoods—agricultural and industrial; however, poor soil and arid climatic conditions hamper agricultural productivity. Bhuj City, nestled in the heartland of Kachchh District, is defined by its semi-desert terrain, the Little Rann salt marsh, and the salt pan. Bhuj City, a commercial hub for wheat, barley, cotton, salt, and cattle, is a major tourist destination due to its salt marsh and is known for handicrafts, including cotton cloth and silver items.

### *Pre-Earthquake Housing Context of Bhuj, Gujarat*

In Bhuj, traditional houses are built predominantly from bio-based materials like mud and thatch. The traditional houses in Bhuj have similar housing typology, including form, materials, and construction techniques. The vast majority of Bhuj's traditional houses are located within the historic walled city, at its core. They contrast with the urban housing that has load-bearing stone masonry (see Fig. 2.4) and reinforced cement concrete (RCC) structures.

### *The 2001 Earthquake Aftermath and the Government's Recovery Policy*

On 26 January, 2001, the Indian state of Gujarat was hit by a 7.9 magnitude earthquake. This earthquake was India's second-most devastating event in over two centuries, since the 1737 Calcutta Earthquake (over 260 years ago) (UNDP, 2001). Nearly 20,000 people died, and over a million homes were damaged or destroyed, causing a $3.5 billion statewide economic loss (UNDP, 2001; World Bank & ADB, 2001).

The epicenter of the earthquake was in Kachchh District (see Fig. 2.2), which suffered 90% of all deaths, 85% of total building asset losses (World Bank & ADB, 2001), and severe damage to handicraft artisans and salt farmers (Balachandran, 2005). Bhuj City, the business center of Kachchh,

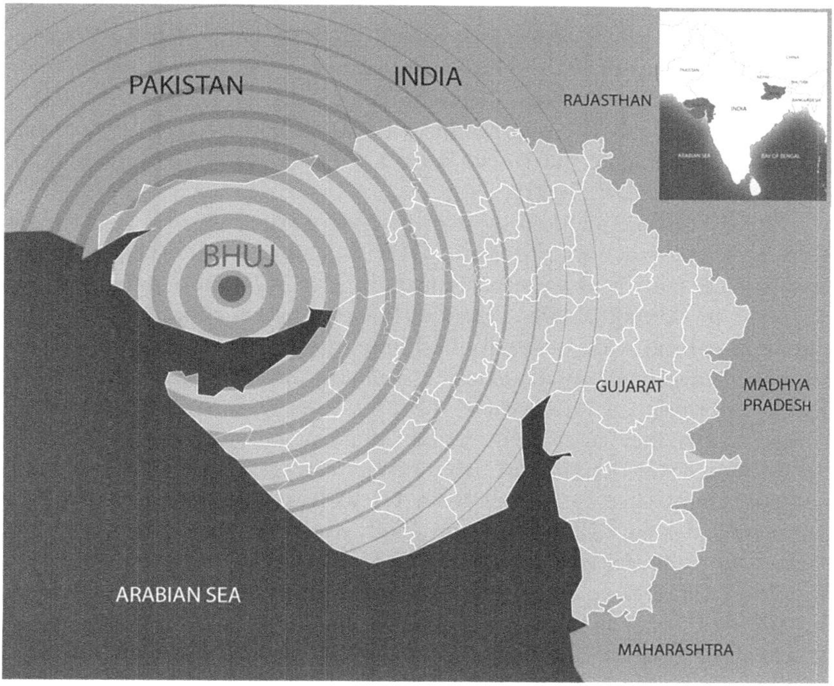


**Fig. 2.2** Map of Gujarat State showing case study location of Bhuj City, affected by the 2001 earthquake (Source: Authors)

witnessed damage to nearly 75% of all houses, or a total of 50,000 houses (BHADA, 2005). Over 40% of all houses destroyed in Bhuj (or nearly 9000 houses) were in the urban core or the walled city. Moreover, 40% of destroyed houses in Bhuj's walled city were occupied by renters (Mukherji, 2008). The reason for large-scale devastation was associated with the old, weak, and multistoried housing stock, dense settlement layout with narrow alleys, and heavy stone masonry construction, which trapped people under its rubble (Balachandran, 2005). The earthquake's impact was compounded for the people of Kachchh as they had experienced six years of drought, two cyclones, and one flood in the decade before the earthquake (UNDP, 2001).

In the wake of the earthquake, the Gujarat State government was spearheaded by Chief Minister Narendra Modi (the current prime minister of India). Modi toppled the then Chief Minister, Keshubhai Patel, to capitalize on this disaster recovery for political consolidation and to get reelected in 2002. The Modi government established the Gujarat State Disaster Management Authority (GSDMA) and launched the Gujarat Emergency Earthquake Reconstruction Programme (GEERP) (GSDMA, 2001b), after borrowing heavily, an amount of $380 million USD, from international institutions like the World Bank and the Asian Development Bank (World Bank & ADB, 2001).

The GEERP was marked by two innovative aspects—an owner-driven housing reconstruction program (ODHR), adopted after government negotiations with civil society organizations (CSOs), and public-private partnerships (Unnati, 2008). The ODHR program is community participatory in nature, with a capacity-building aspect, including the provision of financial assistance, psychosocial assistance, and technical assistance, to disaster survivors to lead their own housing reconstruction process. Financial assistance came from various sources. The government's financial assistance was determined based on the categorization of house damage assessment (five categories from G1 as least damaged to G5 being completely destroyed), house type (high-rise, low-rise, *kachchha*/semi-permanent or *pucca*/engineered), and house location (seismic zone, rural, or urban). CSOs were asked to adopt settlements and were encouraged to provide additional financial assistance of a minimum of 50% top-up to government assistance, as well as provide psychosocial and technical support to communities for housing reconstruction (GSDMA, 2001a). The government's financial assistance was disbursed in three tranches after completion of floor, walls, and roofs, respectively, and certification of compliance from government engineers. The 2001 post-earthquake recovery policy marked a significant departure from the traditional top-down mere financial assistance provision (that typically lead to sub-standard housing outcomes), to an ODHR approach that was community participatory, created space for public-private partnerships, balanced international aid with local capacities, CSO handholding support, and provided a combination of assistance (socioeconomic-technical) and capacity building for compliance with building standards, setting new precedents for India's housing reconstruction and recovery for resilience.

### *Post-Earthquake Housing Reconstruction, GIDC, Bhuj: Collaborative Efforts*

At the city scale, the newly established Bhuj Area Development Authority (BHADA, 2016a) took center stage in guiding reconstruction efforts in Bhuj. While BHADA was provided with a large amount of financial support from the national government for urban recovery, its planning department lacked prior experience in recovery management. Moreover, outdated land use and development plans for the old city (dated 1976) made it challenging for BHADA to navigate recovery efforts (Balachandran, 2005). Hence, BHADA hired a planning consultant firm, Environmental Planning Collaborative (EPC & TCG, 2004) to create accurate maps of the walled city, with residential plots based on land surveys, property data, and damage assessment. EPC's new development plan for the walled city required land readjustment to widen roads for ease of evacuation, as well as building height restriction to two stories for density reduction. The new development plan integrated open spaces for evacuation and revived the traditional water management system, for strategic objectives of earthquake and drought risk reduction (Balachandran, 2005). EPC was applauded for facilitating a participatory planning process in collaboration with a local CSO, the Bhuj Development Council. However, the lengthy and complex town recovery planning process also displaced many people and forced them to live in inadequate, temporary shelters for up to 2 years, and required 5500 households to move out of the walled city due to land readjustment (Mukherji, 2008).

BHADA initially released three government-owned vacant land areas as relocation sites—RTO, Mundra, and Ravalwadi—to accommodate the 5500 households designated to be relocated out of the walled city (see Fig. 2.3). All three sites were peri-urban or located on the outskirts of the

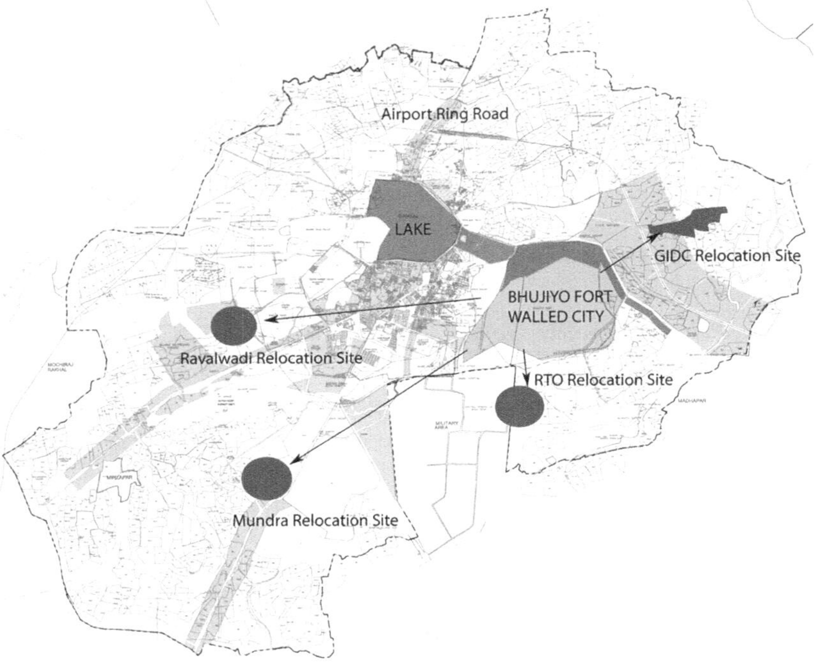


**Fig. 2.3** Relocation sites in Bhuj, Kachchh, after the 2001 Gujarat Earthquake (Source: M. Vahanvati, based on data from BHADA Authority Plan)

walled city. Homeowners could choose a relocation site and buy a land plot (100–200 square meters) at a subsidized fixed rate of 32,000 rupees ($760 USD at that time) (BHADA, 2016b). BHADA provided homeowners additional financial assistance of 3000 rupees ($71 USD) per square meter of built area, with a maximum limit of 150,000 rupees ($3571 USD) per family to rebuild houses, provided their house was severely damaged. Each family could receive a maximum of $4330 USD for land re-purchase and house reconstruction. House owners who were investors and had rented out their properties could receive the same financial assistance for rebuilding damaged housing, provided they ensured pre-disaster renters would reoccupy it (Taheri Tafti & Tomlinson, 2016). At that time, relocation was attractive to many residents due to the larger land size of relocation sites than in the walled city, as well as the promised access to water, sewage, roads, and electricity.

At the state level, the Gujarat State Disaster Management Authority (GSDMA) claimed to have rebuilt over 200,000 houses and repaired over 900,000 houses in 2 years, making it the world's largest and fastest housing reconstruction program (Price & Bhatt, 2009). Almost three-quarters of villages and nearly 87% of all destroyed homes (over 197,000 houses) in Kachchh were rebuilt by their owners under the GEERP program through an owner-driven approach (Barenstein, 2006). For these reasons, GSDMA won the prestigious UN Sasakawa award in 2003 and a Commonwealth award for a paradigm shift in post-disaster recovery, towards mitigation, preparedness, and resilience building (CAPAM, 2004; Vahanvati & Mulligan, 2017).

However, renters and lower socioeconomic groups were forgotten and left out of recovery assistance until three years after the earthquake, living in temporary shelters (see Fig. 2.4). While homeowners in rural and urban areas managed to complete their housing reconstruction in a rapid two years' time, renters or squatters were displaced as they were not included in either land or housing financial assistance packages. The Kutch Nav Nirman Abhiyan (hereafter Abhiyan), a district-wide network of 14 grassroots CSOs and NGOs, kept negotiating with BHADA for the land- and housing-rights of renters and the most vulnerable people. Apart from housing reconstruction, Abhiyan's informal recovery coordination efforts in rural areas through *Setu Kendras* (meaning bridging centers) and the Kachchh Rehabilitation Information Centre (KRIC), with funding support from USAID and UNDP, are noteworthy.

It took nearly three years after the earthquake for BHADA to modify its housing reconstruction program and release a fourth relocation site—Gujarat Industrial Development Corporation (GIDC) (see Fig. 2.3). The GIDC site is a testament to the role of CSOs (Abhiyan)[1] in India, without whom the needs of the most vulnerable people would not have been met. BHADA verbally approved the allocation of 65 square meter plots per family (lower than the 100–200 sq.m. plots in other relocation sites), which would lead

[1] Abhiyan primary members are Anarde Foundation, Ashapura Foundation, Bidada Sarvodaya Trust, Bhimani Khadi Mandal, Gram Swaraj Sangh, Kutch Yuvak Sangh, Kutch Environment and Ecology Cell, Kutch Mahila Vikas Sangathan, Lions Club, Bhuj, Rural Agro Research and Development Society, Sahjeevan, Sarva Sewa Sangh, Shrujan Trust, Vivekanand Research and Training Institute. Associate members are Bhojay Sarvodaya Trust, Kutch Ecology Foundation, Kutch Food Fodder and Forest Development Trust, Kutch Visha Oswal Jain Mahajan, Sahajanand Rural Development Trust, Saraswatam Trust, Vivekanand Gramodhyog Society, Shree Sangh (Ratadiya), Guide (Bhuj), Saath (Rapar), Veerayatan (Bhuj), Corbett Foundation, Yuva (Rapar), Yusuf Maherally Centre (Mundra), Viksat (Bhuj).

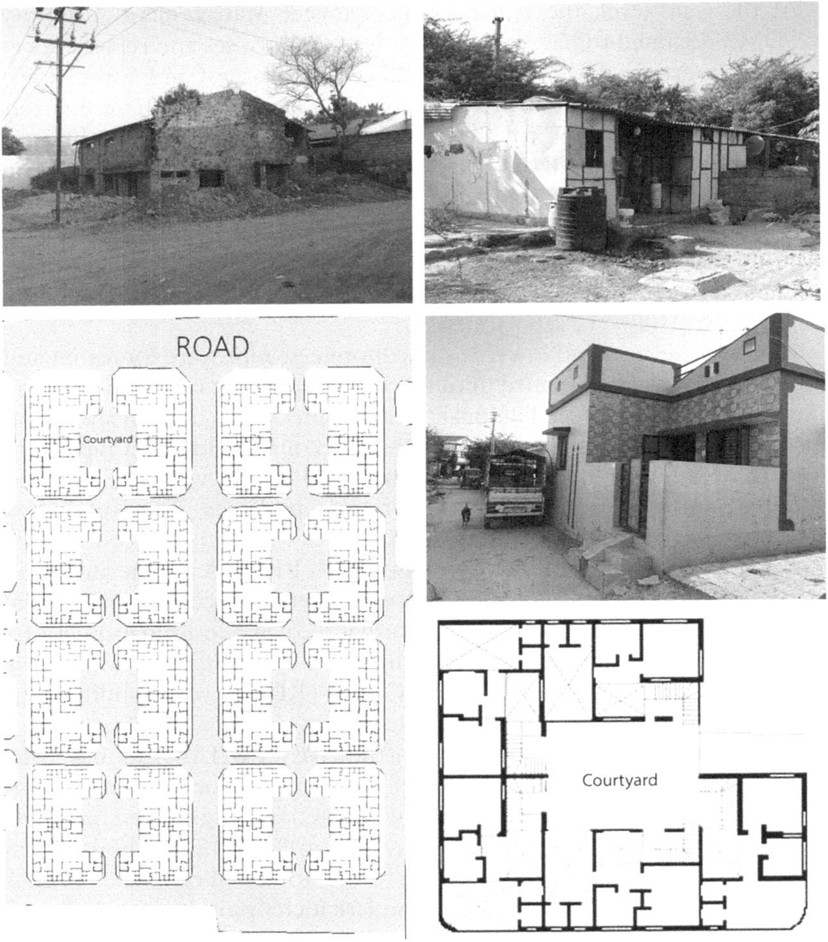


**Fig. 2.4** Rebuilt housing in GIDC relocation site in Bhuj after the 2001 Gujarat Earthquake. Traditional house in Bhuj built using stone masonry (top left); Temporary prefabricated shelters (top right); Rebuilt houses well-adapted and extended over time (20+ years since the 2001 earthquake) (middle right); Sub-cluster of five houses (bottom right) (Source: M. Vahanvati, based on data from Hunnarshala CSO, copyright granted); Eight clusters of 20 houses each, with a central courtyard, totaling 160 houses (bottom left). (Source: M. Vahanvati unless noted otherwise)

to land rights for renters after living there for 25 years. At the time, the GIDC site was already occupied by 1500 temporary prefabricated shelters (Mukherji, 2008) (see Fig. 2.4). For land at GIDC, residents had to pay a subsidized rate of 9000—13,000 rupees ($213—$308 USD) to BHADA, which was much lower than other relocation sites (e.g. 32,000 rupees at Rawalwadi) (interviewee). For housing reconstruction, BHADA urged Abhiyan to provide either fully built houses or pay partially for rebuilding, to make housing reconstruction affordable for low-income communities.

Abhiyan proceeded with settlement layout, housing design, and reconstruction at GIDC. They designed a settlement plan integrating social infrastructure, including community spaces, childcare, post office, bank, temple, mosque; and sustainable initiatives such as sewage recycling, garbage conversion to vermicompost, and recycled waste incorporated into housing reconstruction. They designed a clustered settlement, with each sub-cluster comprising five houses, totaling 160 houses among eight clusters (see Fig. 2.4). For house design, Abhiyan integrated traditional knowledge systems with modern disaster safety principles. The house design was limited to a two-story height, built using stabilized rammed earth, concrete blocks, and RCC structural features like floor-, roof- and lintel-bands, and vertical columns at corners or T-junctions for seismic safety. Through a community-led approach, Abhiyan enabled residents to choose from two housing layouts (each with a room, bedroom, kitchen, and toilet). Initially, Abhiyan allowed residents to choose their immediate neighbors; however, due to social complexities, this was dropped. Rather, caste-based communities were randomly allocated houses next to each other, leading to socially mixed neighborhoods. Abhiyan continued to provide housing recovery support for renters, followed by squatters, from 2003 to 2007, in the following phases.

- **Provided fully-built houses or technical support to renters, to re-build housing on GIDC relocation site (2003 to 2006).**
  - **Phase I**—160 fully built houses were provided by Abhiyan to renters who were relocating from the walled city. These houses were of 32 sq.m. area. Abhiyan fully funded the rebuild, with funding of approximately 292,000 rupees ($6935 USD) for housing and 100,000 rupees ($2375 USD) for infrastructure, from the German KFW bank and HDFC bank (interviewee).

- **Phase II**—48 fully built houses were provided to low-income residents, by Abhiyan, each of 32 sq.m. area.
- **Phase III**—72 households were provided technical support with design and construction supervision by Abhiyan. No financial support for rebuilding housing was provided during this phase.

- **Technical support to squatters to build housing on GIDC relocation site (2007 onwards)**
  - **Phases IV and V**–28 and 150 households, respectively, during Phase IV and V, who were squatters or from slums, or migrants from rural areas, received technical support from Abhiyan, to build their own houses. Residents received funding of 50,000 rupees ($1100 USD) through the government housing policy—*Pradhan Mantri Awas Yogna* (PMAY), which was not part of post-disaster housing recovery funding and/or a bank loan. PMAY required people to build a minimum of 27 sq.m. usable floor area (also termed carpet area in India). The housing construction at GIDC during these phases was overseen by *Kachchh Mahila Vikas Sangathan* (KMVS), a Bhuj-based NGO and part of Abhiyan, who established slum committees, involving women, for collective decision making and maintenance.

Concurrently, with the help of Dr A.S. Arya—the Technical Advisor of GSDMA, Indian Institute of Technology in Bangalore, Auroville in Pondicherry, and National United Nations Volunteer (NUNV) engineers, Abhiyan also instigated legalizing traditional construction technologies and developing guidelines for non-engineered stabilized earth construction (UNDP, 2001).

### *Long-Term Outcomes: 24 Years After the 2001 Earthquake*

As we pass the two-decade mark since the 2001 Gujarat Earthquake, having looked back, we now examine the long-term outcomes of post-disaster housing reconstruction efforts at the GIDC relocation site in Bhuj.

#### *Robustness and Diversity (Buffer Capacities)*

The long-term technical outcomes from housing recovery, 24 years since the 2001 earthquake, are mostly positive. Most houses in GIDC are robust, evident by their withstanding a subsequent earthquake in 2006 measuring 5.6 on the Richter scale. However, residents' confidence in the strength of their houses is tempered by concerns over material choices and construction quality. For example, some residents complained about their walls being built using stabilized earth, without much cement or structural columns and therefore lacking strength. Others were concerned about the perennial threat from floods due to the low-lying settlement (i.e. exposure). These concerns are indicative of a lack of awareness of structural safety features included in the house, potentially because residents were provided with fully built houses.

*Cultural Appropriateness and Adaptability*

The long-term socio-cultural outcomes, including appropriateness and adaptability of rebuilt housing and social mix in the settlement, are mixed. All GIDC residents expressed satisfaction with their settlement's design, house design, and the thermal comfort of their homes. Most residents valued the communal spaces as they allowed for social events such as weddings, fostering social cohesion. However, the social mix of various castes and religions has created tensions and reduced the location's desirability. As noted by one resident:

> "Houses are of excellent quality, but due to uneducated or drunk people who use rough language, the settlement is undesirable and has become like an 'upgraded informal settlement.'" (interviewee)

The impact of several residents' perceptions of lack of safety and reduced desirability was evident, as approximately 40% of houses were not occupied by owners (that is, 7% of houses were rented out and 20% of houses were sold informally) and 12% houses were vacant. Despite the integration of social infrastructure in the GIDC settlement, the social mixing of various castes and religious groups (not of their own choosing) has weakened the social fabric of the settlement, highlighting the critical role community bonds play in resilient recovery.

Views on house maintenance were mostly positive. Some residents mentioned small maintenance costs related to non-structural, superficial cracks in wall plaster from moisture soaking in through stabilized rammed earth walls. Most residents reported major maintenance costs related to the clogging of drainage systems (discussed later in the integration section). Overall, visual inspection showed nearly 60% of houses in good to moderate condition, while 30% of houses were in excellent condition, as evidenced by well-maintained houses with modifications (see Fig. 2.4). Nearly 60% of residents had evidently upgraded amenities, such as house extensions, motorbikes, cars, and air conditioning, suggesting a trajectory of improvement in their living standards and socioeconomic progression.

*Redundancy and Integration*

The long-term functional (access to services and infrastructure) and locational (or exposure) outcomes of housing recovery were mostly negative due to infrastructure deficits. With access to vital services such as safe drinking water and electricity, the residents of GIDC have seen a marked improvement in their quality of life. However, infrastructural deficits include a lack of paved roads, public transport, street lighting, waste management, grocery stores, and health care, highlighting gaps in the integration of essential services with housing. Additionally, most interviewees complained about the clogging of sewage and drainage, which can be attributed to five houses within a sub-cluster sharing one drainage system and a lack of maintenance by government departments. Residents who could afford it have separated their drainage from their neighbors.

Over more than 20 years after the 2001 earthquake, albeit delayed, some basic infrastructure such as paved roads have been built, affecting the everyday lives of the community, their safety and accessibility of livelihood, schools, universities, and healthcare facilities (which are mostly located in the walled city of Bhuj).

*Rights, Equity, and Security*

The long-term equity outcomes from housing recovery are also mixed, mainly due to the lack of timeliness of recovery assistance and ongoing land rights struggles for socioeconomically weaker groups (mainly Muslims, scheduled castes, and renters). Despite government efforts to promote gender equity, evident by the government requiring rebuilt houses to be registered under both spouses' names, India's deeply ingrained social divides based on religion, the caste system, and patriarchal norms, have persisted.

For most residents living in the GIDC site, housing reconstruction has not increased their sense of security, ownership, or stability. Only 15% of interviewees have land rights (or the Redbook of land title). Anecdotal evidence suggests that BHADA's mismanagement of land titles have led to residents receiving letters about growing interest on non-payment of initial land title documentation fees, as noted by one resident:

*"In 2004 we were supposed to pay 9,750–10,000 rupees for land title to get the Redbook from BHADA. By 2012, BHADA demanded 6,000 rupees more from us as an interest on non-payment, which became 18,000 rupees by 2018–19. People who had paid 6,000 rupees also received an interest notice. Seems like BHADA didn't process the money. [Now, in 2022] we have to pay 28,000—29,000 rupees."* (interviewee)

For low-income residents who could not afford the large sums of the payments, acquiring land title ownership has caused a lot of anxiety. As a result, despite homeowners being mandated to not sell their land and houses for 25 years from reconstruction, many residents have sold their houses informally. However, those who have bought houses informally have ended up paying twice—once to the previous owner, and again to BHADA. Moreover, they have three informal documents for land rights, including (1) a receipt of the informal house sale, (2) power of attorney from the previous owner, and (3) a land title document with the name of the previous owner. Growing informality is also reflected in the property values at GIDC, as noted by a resident:

"*GIDC houses are valued at 2–3 lakh, while RTO houses are valued at 50 lakhs.*" (interviewee)

The situation of the GIDC site underscores the importance of land- and housing-rights for all residents, as it hinders more inclusive and equitable recovery (UNDP, 2001).

*Resourcefulness (Adaptive Capacities)*
The long-term resourcefulness or capacities outcomes from housing recovery are mixed, with increased institutional and NGO capacities, but mixed levels of improvement in community capacities. Visual observations suggest improvements and noticeable incorporation of structural safety features in most house extensions. Potentially, frequent changes in homeowners from reselling or renting is depleting residents' risk awareness, evident from unnecessary concerns about the safety of their house due to superficial cracks. The capacity of government agencies (GSDMA at the national level and BHADA at the local level) has been notably improved, evident in strategic recovery planning, land readjustment as part of urban development plans for the walled city of Bhuj in efforts to reduce future risk, new development controls, rural development, revised building

regulations, and the strict enforcement of hazard-safe building standards. BHADA, despite questions around lack of prior urban recovery experience and equity, particularly for renters, managed to improvise and create a significant shift towards more inclusive urban development. The continued presence and operation of GSDMA and BHADA to date is potentially fostering a culture of continuous learning and resilience building.

#### *Summary of Long-Term Outcomes and Key Factors of Housing Reconstruction in GIDC Relocation Site, Bhuj, Gujarat, that Led to Resilient Outcomes*

While in the short term, renters and socioeconomically weaker groups were left out of recovery assistance, in the longer term (24 years after the 2001 Gujarat Earthquake), analysis at the urban, relocation site of GIDC demonstrated that housing reconstruction improved residents' living quality, despite a few remaining challenges to overcome. Without equal land rights, integration of infrastructure, and social cohesion, community resilience outcomes are hindered.

Looking back, key factors during housing reconstruction post Gujarat Earthquake that helped catalyze resilient outcomes in the long term are (i) government good-will, making space for CSOs and the private sector, and agility to account for the needs of lower socioeconomic groups who were previously neglected in recovery assistance, upon negotiations with CSOs; (ii) an owner-driven housing reconstruction program that focused on strategic disaster risk reduction, as well as capabilities and capacities strengthening of residents; and (iii) the important role of CSOs in translation of policy to practice. However, the challenges of non-compliance with building standards, and deep-rooted issues related to the caste system, have crept back in with the potential to jeopardize the sustainability of such interventions.

### Case 2: Bihar Housing Reconstruction Post-2008 Kosi River Floods

Bihar is nestled in the eastern part of India in the Indo-Gangetic plains, bound by the Himalayas and Nepal to the north and Bangladesh to the east. Bihar's vulnerability is linked to its unique geographical location, climatic condition, extreme socioeconomic weakness, and government capacity deficit. Historically, the state was known to be India's spiritual hub (being the place of enlightenment of Lord Buddha and Lord Mahavir's birth place), cradle of knowledge (being home to the world's first universities—Nalanda and Vikramshila in the fifth and eighth centuries BC, respectively, and the place where ancient texts like *Archashastra*—science of economics were created), and a political hub (being the world's first democracy) (UNDP, 2014). Historically, colonial exploitation and seven famines, including the deadly 1943 famine that caused three to four million deaths, have had lasting damage on the state. Presently, the state is only known for its spiritually meaningful past but is marked by extreme poverty, inadequate housing, caste-ridden social order, non-existent rule of law, being prone to Naxalite terrorist attacks, weak governance, and is one of the least-growing economies in India.

Geo-climatically, the state is marked by its changing hydrology and heavy (often erratic) monsoon rains that occur on an annual basis, as it is in a humid sub-tropical temperate zone. The plateaued landmass, situated at the foothills of the steep Himalayan Range, is constantly shaped and reshaped by its 21 rivers changing their courses. The state's geo-climatic conditions make it prone to hazards including floods, storms, and earthquakes. The River Ganges splits Bihar into two parts—north and south. Most rivers flow north–south, draining into the River Ganges. Floods affect nearly 73% of Bihar's landmass and earthquakes affect 28 of 38 districts, most of which are located in earthquake zones VI–V (zone V being the highest earthquake risk zone) (FMIS & GoB, 2009). Flowing south, the Kosi River's destructive meanders have earned it the title of "the sorrow of Bihar". Over the last 120 years, the Kosi River has changed course every 50 to 100 years, carving 12 new channels and moving 113 kilometers west (Meena et al., 2014). Due to climate change and receding Himalayan glaciers, the interval between highly devastating floods—in 1978, 1987, 1998, 2004, 2007, and 2008—has reduced from 9 years to every year (Pritchard & Thielemans, 2014). The key government strategy for flood mitigation has been technological, or to build embankments along the Kosi River. Embankments increased from 160 km in 1954 to nearly 3400 km by 2008 (Pritchard & Thielemans, 2014). Despite its flood-prone landscape and socioeconomic and political deficits, Bihar's rich water resources and fertile land that allows three crops a year attract people to live with flood risk, as part of Bihar's predominantly agrarian economy.

### *Pre-flood Housing Context*

Most residents in Bihar continue to live in rural areas. Housing in Bihar is a blend of traditional bio-based homes or *kutcha* houses (impermanent houses) and *pucca* houses (permanent houses). Traditional houses are typically crafted from local and natural materials like mud, bamboo, or wattle and daub (see Fig. 2.6), reflecting an ephemeral relationship of people with their environment. Pucca houses are built using industrial materials like brick, concrete, or corrugated sheet metal. Traditional houses enable easy relocation in the event of floods. For example, in the event of floods, flood waters wash away the mud plaster of panelized woven bamboo mat walls, which being lightweight allows residents to transport housing structures and materials on boats and relocate to higher ground on land unaffected by floods. This traditional housing culture, developed by the region's artisans, skilled over many generations, now faces the test of time. For example, untreated bamboo used for traditional housing construction tends to get infested with borers if not treated or if built without lifting the bamboo above ground to keep away from groundwater, reduces the building's lifespan, and requires ongoing maintenance or replacement every few years, which is now challenging due to changing lifestyles and livelihoods (Vahanvati, 2018b).

### *The 2008 Flood Impact and Government Recovery Initiatives*

In August 2008, the Kosi River floods displaced over 3 million people (UNDP, 2009), damaged 236,632 houses, and caused socioeconomic destruction in about 1000 villages (GoB et al., 2010; GoI & UNDP, 2011). The districts of Supaul, Madhepura, Araria, Saharsa, and Purnia, were among the worst affected. The estimated housing loss was valued at approximately 8.8 million rupees ($195 million USD) (UNDP, 2009, p. vii). The floods became devastating because of a breach in Kosi River's embankment in Nepal, causing the river to change course, shifting eastward and inundating areas that had not witnessed floods for over 50 years (GoB et al., 2010) (see Fig. 2.5). This event underscores the inadequacy of existing water infrastructure, predominantly reliant on technical engineering solutions for flood mitigation.

Supaul district, due to its dense population of approximately 1.73 million people as per the 2001 Census, was the worst hit due to its proximity to the border with Nepal where the embankment breach occurred. In

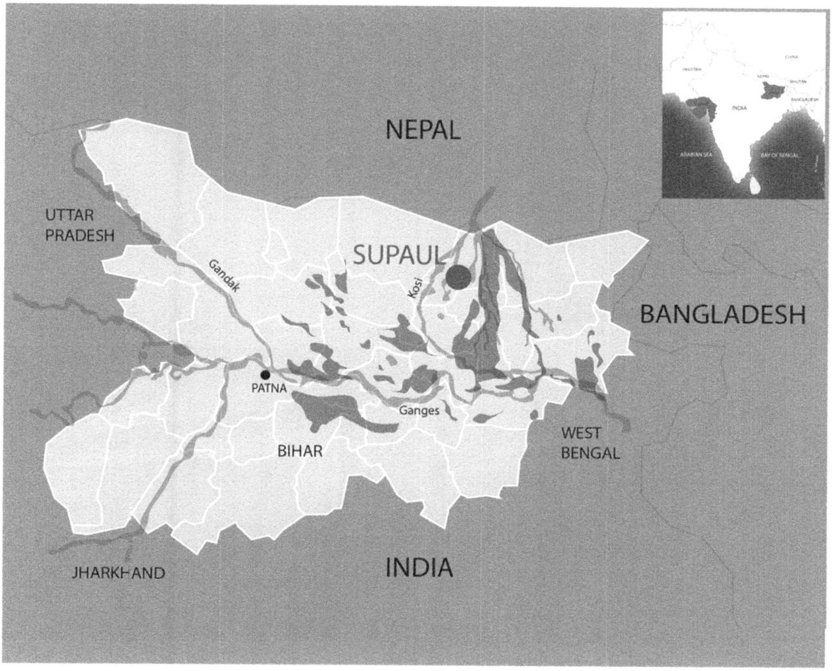


Fig. 2.5 Map of Bihar showing the case study location in Supaul, affected by the 2008 Kosi River floods (the darker blue shows change in Kosi River's course, causing flooding) (Source: Authors)

Maanganj settlement in Tribeniganj Block in the district of Supaul, the floods damaged approximately 60,000 *kutcha* houses (GoB et al., 2010), owned by predominantly lower socioeconomic groups, which make up roughly three out of five households.

### *Post-flood Housing Reconstruction, Maanganj, Supaul, Bihar: A Two-Phase Approach*

Since the Bihar State Disaster Management Authority (BSDMA) was established just eight months before the 2008 floods (BSDMA, 2012), they had no prior recovery experience, limited governance capacity, and there were concerns about corruption (ODR collaborative, 2008). Despite

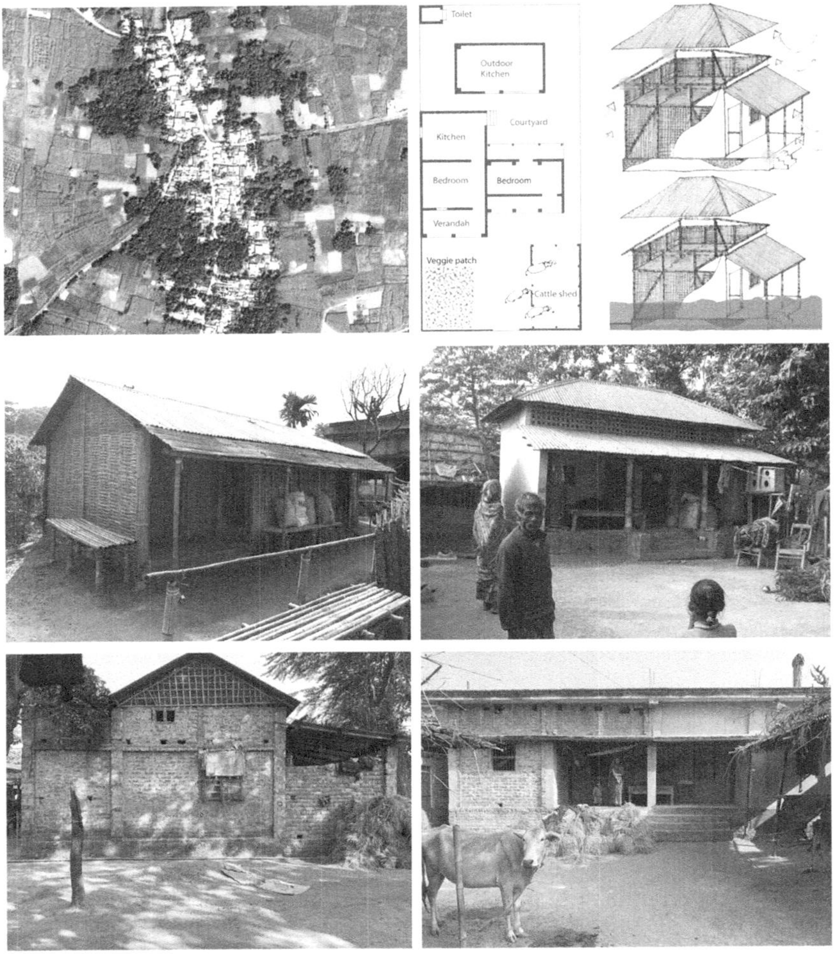


**Fig. 2.6** Housing reconstruction, in-situ and owner-driven, in Maanganj village in Supaul district of Bihar, after the 2008 Kosi River floods. Plan and 3D drawing of a cyclone, flood, and earthquake safe (multi-hazards safe) house design for Bihar recovery, proposed by ODRC, using local material such as bamboo (top right) (Source: Hunnarshala, copyright permission granted); Typical traditional shelter in northern Bihar (middle left); Rebuilt house, post-2008 Kosi River floods using treated bamboo, CGI sheet roofing and pile footing (middle right); Extension of rebuilt houses, 17 years since floods (bottom left and right). (Source: M. Vahanvati unless noted otherwise)

the challenges, the then Chief Minister—Nitish Kumar—aspired to a strategic recovery plan as opposed to the prior nominal financial aid provision, which had failed to improve the housing quality or mitigate flood devastation. The BSDMA thus invited the Owner-Driven Reconstruction Collaborative (ODRC) to help build state government's capacities in recovery policy development, in-field coordination, and housing reconstruction implementation. The ODRC was a South Asia Regional collective of CSOs including the Kutch-based CSO Abhiyan, Indian government authorities including the GSDMA, and international think tanks with experience and expertise in owner-driven housing reconstruction. The ODRC played a pivotal role in formulating the Kosi Reconstruction and Rehabilitation Program (KRRP) (GoB & ODRC, 2008b) that adopted an owner-driven housing reconstruction approach. For implementation, ODRC proposed a two-phase recovery approach, with Phase I involving piloting the KRRP program in two villages to build government capacity at all levels (for detailed discussion on these two sites, see Vahanvati & Beza, 2017) and Phase II involving ODRC handing over the project to BSDMA to scale up housing reconstruction efforts.

*Phase I Housing Reconstruction in-Situ, in Two Pilot Villages (2008–2011)*

The ODRC led Phase I piloting of housing reconstruction in two villages—Orlaha and Puraini. Built upon the experience of *Setu Kendra's* success in Kachchh, post 2001-Gujarat Earthquake, ODRC set up *Kosi Setu Kendras* (KSKs) as a decentralized governance mechanism to coordinate recovery efforts. *Kosi Setu Kendras* acted as one-stop shops for residents, providing them with socio-technical support for setting up bank accounts and facilitating transactions, providing skills training to bamboo artisans and brick masons, and monitoring on-site owner-driven building (ODR collaborative, 2008). Demonstration houses were built to show villagers key structural safety features for earthquake, cyclone, and flood safety. Technical guidelines were developed with the help of Professor Arya, the Indian National Seismic Advisor, for local materials such as bamboo and bricks. Additionally, skills training was provided to local artisans and residents (ODR collaborative, 2010). *Kosi Setu Kendras* embodied a paradigm shift in becoming institutionalised, and empowering residents or reviving their confidence in their capacity to rebuild robust houses.

Beneficiaries during Phase I were selected by the block-level government officials or the village chief, and the list of their names was displayed publicly in the village square for transparency and dispute resolution. Land rights resolution took much longer because 30% of affected people were landless. Settlement-based communities were encouraged to resolve land rights issues within their community and were informed that housing reconstruction could not start before land rights resolution. Subsequently, the community identified land parcels that they were happy to sell to the landless, and the Bihar government provided those land parcels to the landless at a subsidized rate (ODR collaborative, 2008). The assistance package for housing reconstruction during Phase I included financial assistance of 55,000 rupees ($1220 USD) for households to rebuild a core house of 15 square meters, and socio-technical support for integrated structural safety features. The money was disbursed in two tranches (when houses were built up to lintel level and on completion of core shelter), upon compliance certification by local artisans (GoI & UNDP, 2011). Residents had the freedom to design their own house, and choose materials and methods of construction labor, but compliance with safety features in housing was mandatory. To incentivize speedy completion within two years, an additional 5000 rupees ($110 USD) and 2300 rupees ($50 USD) were provided for solar lighting and toilet construction, respectively (GoB & ODRC, 2008b; GoB & World Bank, 2010).

Such piloting before policymaking by ODRC helped finetune the housing recovery package, participatory processes, technical guidelines, and institutional mechanisms, as well as build government capacity to manage recovery. Such an innovative approach, involving piloting before policy making, can be termed an "agile" approach, which has rarely been witnessed in post-disaster housing reconstruction.

By the end of 2010 (two years after the floods), only 130 houses were rebuilt by residents, in two pilot settlements in the Supaul district (Vahanvati & Beza, 2017; Vahanvati & Mulligan, 2017). Apart from resilient house reconstruction, these settlements gained access to electricity (solar lighting), waterless toilets, and street lighting, for the first time in residents' lives. However, the rest of Bihar's population, approximately 998 villages or 100,000 households, were still awaiting recovery assistance for over two years.

#### *Phase II Housing Reconstruction, in-Situ, Up-scaled to All Flood-Affected Communities (2010–2018)*

Phase II was envisaged as ODRC handing the recovery program over to BSDMA for scaling up housing reconstruction for approximately 100,000 families. However, for unknown reasons, in 2010 ODRC left Bihar, after an abrupt handover to BSDMA, leaving BSDMA to manage recovery efforts on their own without any capacity-building support (GoB & ODRC, 2008a). To manage the mammoth task of recovery, BSDMA borrowed $52.2 million USD from the World Bank (2018, p. 4). The recovery program was renamed from KRRP to *Bihar Aapada Punarwas Evam Punarnirman Society* (BAPEPS) (translated as Bihar Disaster Rehabilitation and Recovery Program), potentially to distinguish the two phases.

While the *Kosi Setu Kendras* continued for coordination of housing reconstruction during Phase II, they underwent minor changes. Each *Kosi Setu Kendra* (KSK) assisted roughly 4000 houses (World Bank & GFDRR, 2013) and was comprised of a multidisciplinary team. Each KSK included the chief program manager (one per block, the same in both phases) at the sub-district or block level, a local engineer (reduced in Phase II to one engineer for every 1000 houses in each village), an accountant (one per block, introduced in Phase II), an information technology operator (one per block, same in both phases), and social coordinators (reduced from one per 500 houses in Phase I to one per 1200 houses in Phase II).

All *Kosi Setu Kendra* members were appointed and paid by the Bihar government. The number of engineers and social coordinators were reduced in Phase II compared to Phase I, reducing socio-technical support they could provide to disaster-affected residents. Moreover, the local *Raj Mistrys* (translated as Master Carpenters) who were appointed instead of engineers during Phase I, were not re-appointed during Phase II to supervise and check compliance of rebuilt houses with building standards, because of their lack of engineering degrees. These changes represent the World Bank bringing in the Western system and rewarding those who had such educational qualifications, as opposed to the local, traditional, and indigenous knowledge systems. While the social-technical support was reduced, the financial assistance package for residents was increased from 55,000 rupees ($1220 USD) in 2008 to 85,000 rupees ($1930 USD) by 2016, to keep up with the hikes in material prices, which the government neither subsidized nor controlled well, as noted by a resident:

> "The government money was not sufficient, as material prices kept increasing e.g. price for 1000 bricks in 2011 was 4000 rupees, and 12,000 rupees in 2022; mason's labour charges were 400 rupees per day in 2011, whereas 600 per day in 2022. Building a house cost 150,000 rupees ($3,405 USD) in 2011, now it costs 400,000 rupees ($9,080 USD) [that is, a three-fold increase]." (interviewee)

Financially, Phase II brought additional complexities. Most residents living outside of the two pilot settlements opted to get funding from the national government's rural housing scheme—*Indira Awas Yogana* (IAY) (now termed *Pradhan Mantri Awas Yogna* or PMAY) in the 22 months that passed before Phase II began, receiving 45,000 rupees ($1020 USD) to rebuild their homes. Hence, Phase II reconstruction funding had to be improvised to be able to provide these residents with the difference between IAY (or PMAY) and BSDMA recovery funding, amounting to $410 USD (GoB et al., 2010, p. 19).

Maanganj village—the case study settlement and one of the 64 villages in Supaul district—which was part of the Phase II BAPEPS program, had delays in housing reconstruction. Their residents completed housing reconstruction or rehabilitation by 2018 (nearly 10 years after the floods). Three rebuilt housing typologies were evident in Maanganj: (i) brick walls with reinforced cement concrete (RCC) structure and flat-terraced roof; (ii) brick walls with RCC structure and gable roof with corrugated galvanized iron (CGI) sheet roofing; and (iii) treated bamboo structure with woven bamboo mat and cement-stabilized mud plaster walls and CGI sheet roofing. The majority of houses in Maanganj were built using brick masonry, accompanied with outhouses or animal shelters that were built using bamboo (author observations, see Fig. 2.6).

Beneficiaries during Phase II were selected by the block-level government officials; however, transparency or dispute resolution during this phase is questionable because some discrepancies emerged in beneficiary lists. According to the village chief, all the 2200 households in the village had received assistance and had recovered their housing. However, the Census of India (2011) had recorded 2607 families in Maanganj village. Upon corroborating this data with residents during interviews, it emerged that approximately 300 families had missed out on assistance for housing reconstruction. When re-examined against the caste of those who missed out on assistance (or non-beneficiaries), results highlighted that most non-beneficiaries were from the scheduled caste (SC), scheduled tribe (ST), or other backward class (OBC), highlighting socio-economic bias in beneficiary selection.

It seems apparent that BSDMA struggled to complete the reconstruction program, evident in the World Bank having to extend the Phase II completion deadline several times. The World Bank announced the project closed in 2018, despite the completion of only 56,758 houses out of 100,000 houses (or 56%) and building of 44,946 toilets, noting the program to be "moderately successful" (World Bank, 2018, p. 33).

### *Long-Term Outcomes: 17 Years Post-2008 Kosi River Floods*

*Robustness and Diversity (Buffer Capacities)*
The long-term technical outcomes from housing recovery, 17 years since the 2008 Kosi River floods, are mixed. The robustness of most rebuilt houses in Maanganj village, during Phase II, is evident by them withstanding a subsequent earthquake of 6 on the Richter scale in 2015, high winds in 2020, and ongoing annual floods (interviewee). Firstly, over time, the design of houses has improved—to be lifted off the ground, use treated bamboo or brick include pile footings, and have an attic space. However, inconsistencies in construction quality and non-adherence to building standards have emerged. Poor quality construction was evident in exposed reinforcement bars in RCC beams, corroding reinforcement bars, and anecdotal evidence of residents using fewer metal ties in RCC structures to save costs. Secondly, anecdotal evidence highlights increased corruption as residents bribed the engineers from *Kosi Setu Kendra* to sign off on compliance certificates, despite poor construction quality in order to get the next financial instalment. Potentially, a decline in construction quality and compromise in building compliance may have resulted from a combination of the pressure for housing reconstruction completion to adhere to timelines set by the World Bank, and/or have something to do with a lack of accountability and increasing corruption among government engineers. It raises critical concerns about Bihar's housing culture slipping back to pre-disaster construction standards and government officials resorting to corruption. Over half of the houses in Maanganj village remain unfinished since housing construction is typically incremental among lower economic groups. Lastly, limited land availability has meant no reduction in exposure to floods or earthquakes.

*Cultural Appropriateness and Adaptability*
Long-term socio-cultural outcomes from housing recovery, in terms of the cultural appropriateness of housing, its adaptability over time, and social mix in Maanganj village are mostly positive. Almost all rebuilt houses are occupied and have been personalized, reflecting residents' cultural preferences. Most residents were satisfied with the design and thermal comfort of their houses. Over 40% of residents had extended their homes vertically or horizontally, revealing a propensity for adaptability. High satisfaction rates are related to residents' freedom to design and build their houses as per their own values and preferences. However, most residents were dissatisfied with the size of their house (215 square feet or 19 square

meters), potentially because they were unaware of the fact that the government's funding assistance was for rebuilding a core house, not an entire house.

Overall, only one construction technology—brick and corrugated metal sheet/RCC-based—has taken root in Maanganj. There is limited or no use of traditional, bamboo-based construction, despite CSOs encouraging its use and having developed technical guidelines for this traditional and green construction material (Vahanvati, 2015). The house design has also changed from a pitched roof to a flat terrace roof. Over 40% residents in Maanganj village had chosen corrugated galvanized iron (CGI) sheet roofing and desired RCC terraced roofing, despite those materials being climatically and thermally uncomfortable. The reasons for such materials selection were high social status, hazard stability, and perception of thermal comfort during hot and humid summer months (see Fig. 2.6). Such evolving housing typology and construction technology, marked by a departure from biodegradable materials like mud and bamboo, mirrors the shift in the economic landscape, where cyclical urban migration for livelihoods and increasing wealth is reshaping people's aspirations for their rural habitats.

*Redundancy and Integration*

The long-term hazard exposure and functional outcomes from housing recovery are mostly positive due to access to basic infrastructure and services, previously unavailable to most residents in northern Bihar. By 2018, significant progress was made in the rural village of Maanganj, improving the villagers' access to basic services, evident by 80% of families having access to toilets or 44,946 toilets constructed (World Bank, 2018, p. 33). However, cultural preferences for water-based toilets have resulted in the underutilization of waterless toilets. Most families who could afford it have replaced waterless toilets with water-based ones, along with septic tanks. Additionally, practical challenges of maintaining solar lighting along streets have resulted in several of them falling into disrepair. While upgraded village infrastructure is commendable, such development opportunities in rural areas are also limited by challenges of longevity and residents' inability to maintain them.

Overall, the quality of life of all residents has improved due to access to paved roads and solar streetlights, solar electricity (unavailable before the 2008 floods), safe drinking water, water tanks, toilets, and better housing. Moreover, durable and stronger houses requiring lower maintenance have led to financial savings for most residents, which they have invested in diversifying livelihoods, such as buying motorbikes, and sending their kids to schools instead of agricultural work. As noted by one resident:

> "Every two-three years we needed 15,000–25,000 rupees to rebuild our bamboo house. Now, there's little need for maintenance, requiring only approximately 2,000–3,000 rupees every two years, helping us save a lot."

Such improvements in living standards does prove co-relationship but does not necessarily prove causational links to housing reconstruction. However, evidence of poverty and low standards of living among non-beneficiaries suggest potential links between living standards with the robustness of rebuilt housing.

Redundancies were integrated into housing reconstruction, allowing walls [not structure] at ground level, below the lifted floor, to give way during floods. It is unclear if such redundancies were incorporated at the settlement scale or macro-scale of Kosi River's water catchment area for flood risk reduction. While the Bihar government has been working on an integrated Kosi River catchment management strategy with funding support from the World Bank, it remains unclear if they have integrated green infrastructure along with grey infrastructure like embankments, along Kosi River to mitigate floods. Moreover, examining nature-based solutions for flood mitigation or their effectiveness is beyond the scope of this research.

*Rights, Equity, and Security*

The long-term equity outcomes from housing recovery are mixed, casting light on the stark reality of socio-political nature of recovery. Empirical evidence proves that approximately 300 households or 8–13% of Maanganj village's population missed out on assistance for housing reconstruction, and these households were from the weakest socioeconomic group (i.e. scheduled caste and extremely backward class). Their omission points to a troubling undercurrent of corruption and impacts of power relations between residents and the village chief (i.e. Mukhiya), as noted by one interviewee:

> "Social power or capital is a double-edged sword. The positive is that those who have it recover faster. The negative is that they [in position of power

or higher social capital] feel invincible, above the law and untouchable by rules, regulations, and compliance. They think they can buy their way out of anything, often benefiting from floods." (interviewee)

As noted, a lack of social power can cause deprivation, while too much power can tempt a person to engage in corruption. Power-based discrimination has led to inequities in recovery, often reinforcing vulnerability. Anecdotal evidence from a few residents suggest that they had to pay bribes to the *Setu Kendra Hub*'s engineers for a compliance certificate to get the next financial instalment, signaling rampant corruption during Phase II reconstruction. Such deprivation can have a ratcheting effect on the poor, as aptly noted by one non-beneficiary.

"A good house is important for me to feel happy. If one doesn't have a house for their children, it's troubling and reinforcing poverty." (interviewee)

Equity issues did not emerge in securing land rights post flood in Maanganj village because all households had land titles pre-floods. The government demanded a nominal payment from residents for reissuing lost land title papers post flood. Moreover, as noted earlier, the government also helped landless residents with the provision of land at a subsidized rate. These findings reinforce the importance of land rights, as well as adequate, robust, and low-maintenance housing compared to pre-flood housing, as foundational for residents' improved sense of security and resilient recovery.

*Resourcefulness (Adaptive Capacities)*

The long-term capacities or resourcefulness outcomes from housing recovery are mixed because of some improvement in community awareness but a lack of evidence of any such improvements in government capacities. Empirical evidence from interviews and on-site observations in Maanganj village affirmed an increase in community awareness about safety features in housing construction and the ability to make risk-informed decisions. This is also evident from one interviewee's statement.

"The biggest improvement in housing construction is the inclusion of safety features like deep foundations, raised floors (above the flood level), and attic space. I think we need a minimum three feet deep foundation (though the minimum required is five feet deep); four rods in RCC and one plinth beam is essential. However, three beams'—plinth, lintel, and roof—levels are important."

While residents' awareness about safety features in housing has increased, their material preference has changed, highlighting the need for sustained capacity building in the construction sector. The community's social bonding remains high. However, it is unclear if the self-help groups established during recovery are still active to be able to voice their concerns.

Due to limited evidence, it is unclear whether the government capacities for housing reconstruction or flood mitigation have improved, considering the recovery efforts took more than 10 years. Moreover, the World Bank report rated the reconstruction program in Bihar as a failure in 2011, and later revised it as moderately satisfactory.

### *Summary of Longer-Term Outcomes of Housing Reconstruction and Key Factors That Led to Resilience Outcomes in Maanganj Site, Supaul, Bihar*

Once again, like in Gujarat, the majority of flood survivors (most of whom are also socioeconomically weaker groups) of Bihar were left without any recovery assistance for nearly two years post floods, while BSDMA built their capacity for housing recovery management with the help of ODRC, through piloting in two villages (Vahanvati, 2018a). Recovery for residents in Maanganj village was protracted for over 10 years, since many households initially rebuilt their homes with funding from IAY (now PMAY), which did not mandate compliance with building standards, and later rebuilt their homes again, to better building standards. For residents of Maanganj who have chosen to live with floods, building and rebuilding housing has become the only constant. For those who rebuilt housing as part of Phase II in Bihar's Maanganj village, the houses are robust, even 17 years after the devastating 2008 Kosi River floods. However, incrementally, non-compliance with building standards is slipping back into the housing culture. Residents' socioeconomic quality of life has improved due to land rights and access to basic services and infrastructure; however, their exposure to floods has not been reduced.

In terms of capacities, while the government of Bihar took a leap towards developing a strategic recovery plan post flood, adopting agility for fine-tuning owner-driven housing reconstruction program, which was adopted for the first time in Bihar's history, it is unclear whether the government's capacities for disaster recovery management improved. This is partially due to the absence of capacity-building support from ODRC after two years, during the scaling up phase. Evidence suggests government capacities may not have improved due to the recovery program taking longer than 10 years. Contrary to capacities for resilience strengthening, corruption seems to have crept in again, threatening a return to traditional financial aid approaches to recovery.

## Comparative Analysis of Two Case Studies from India

In comparing housing recovery policies in Gujarat and Bihar, at the state level, 24 years after the 2001 Gujarat Earthquake and 17 years after the 2008 Kosi River floods in Bihar, a few similarities, differences, and changes were identified. There was a strong similarity in recovery policies between the two Indian states of Gujarat and Bihar, notably the active role of CSOs in shaping recovery policies. Other similarities include community capacities and an resident agency-centered approach (termed as an owner-driven housing reconstruction approach), with financing from various sources, due to India being a lower-middle-income country, and a focus on strategic objectives of disaster risk reduction along with addressing the deep-rooted socio-technical challenges. However, differences were also evident from 2001 to 2008 in terms of changes in governance and coordination mechanisms from centralized to decentralized; in reconstruction program implementation shifting from public-private partnership to a highly collaborative yet government-led; and in inclusion of land rights for the poor or landless from early on, due to the application of lessons learned from 2001 to 2008.

We now focus on the long-term outcomes of the implementation of housing reconstruction projects (or policy translation into practice) in the two Indian case studies—GIDC, Bhuj, nearly 24 years after the Gujarat Earthquake and Maanganj, Supaul, nearly 17 years since the Kosi River floods in Bihar—to identify similarities, differences, changes and adoption of lessons learned (see Table 2.1). Such comparative analysis of long-term outcomes allows us to explore key factors as drivers that catalyzed resilience outcomes.

### *Similarities, Differences, and Changes in Housing Reconstruction Projects in India, over Time*

#### *Robustness and Diversity (Buffer Capacities)*

Despite India's predominantly informal building and construction sector, reconstruction brought radical changes in both cases in the short-to-medium term, but such robustness is slipping back to pre-disaster standards in the long term (see Table 2.1). Such positive results in the short-to-medium term can be attributed to skills trainings and financial assistance tied to building standards' compliance, as part of the owner-driven housing reconstruction (ODHR) program by the state governments and the active role of CSOs providing socio-technical support to disaster-affected communities for rebuilding their homes. Moreover, building standards were revised and traditional technologies were included in the building code. However, without sustained capacity-building support by the state government for the construction labor to attain certification and embed a culture of safe construction in ongoing development activities, housing robustness is slipping away. Exception are a few projects in Gujarat where CSOs encouraged trained masons to get certification in seismic-safe construction.

*Cultural Appropriateness and Adaptability*
The long-term socio-cultural outcomes from housing recovery are mixed in both case studies. While higher satisfaction with the house design and its adaptability was fairly similar in both case studies in Bihar and Gujarat, lower satisfaction among GIDC communities (in Gujarat) due to an undesired social mix of neighbors highlighted the deep-rooted challenges associated with the caste system in India. In communities with higher satisfaction, the adaptability of housing is related to the adoption of the ODHR program by the state governments and partnership with CSOs who created an enabling environment for communities to make risk-informed decisions.

*Redundancy and Integration*
The long-term services and infrastructure integration outcomes from housing recovery were negative in GIDC's urban relocation site, post 2001 Gujarat Earthquake, due to infrastructure deficits. However, such long-term outcomes were mostly positive in Bihar's Maanganj village's in-situ reconstruction project post-2008 Kosi River floods due to access to basic services and infrastructure for the first time in residents' lives. In both cases, residents' exposure to hazards was not reduced due to limited land availability. Such outcomes can be related back to key factors such as the local government's inadequate capacities, power dynamics, or corruption, such as BHADA's inability to meet the road or drainage infrastructure needs of the GIDC community in Gujarat in a timely manner, and the block-level government's corruption in Bihar leading to

**Table 2.1** Comparing resilience characteristics as long-term outcomes of two housing reconstruction projects in India. (●●● = strongly agree; ◎ = agree; ± = neutral; ⊠ = disagree; ⊠⊠⊠ = strongly disagree) (Source: M. Vahanvati)

| *Resilience characteristics* | *Resilience outcomes in the housing systems over the long term* | *Description of long-term resilience outcomes in housing, communities, and institutions* | *GIDC Bhuj 2001 Gujarat EQ (urban)* | *Manganj 2008 Bihar floods (rural)* |
|---|---|---|---|---|
| Robustness and diversity (buffer capacities) | Planning and building standards | Well located for reduced exposure | ⊠⊠⊠ | ⊠⊠⊠ |
| | | Building standards stronger/revised | ●●● | ●●● |
| | Structural safety | Structural hazard-safety features explicitly integrated in housing | ●●● | ◎ |
| | | House withstood another disruption | ●●● | ●●● |
| | | Good construction quality | ●●● | ◎ |
| | | Choice of materials and construction technologies | ●●● | ●●● |
| | | Traditional/local construction labor incorporated | ●●● | ●●● |
| | | Replication of improved construction systems | ⊠ | ◎ |
| Cultural appropriateness and adaptability | Housing design | Design with diverse housing typologies | ◎ | ●●● |
| | | Culturally responsive spatial layout | ●●● | ●●● |
| | | Thermally comfortable design | ●●● | ●●● |
| | | Design flexibility for extension or adaptation | ◎ | ●●● |
| | | Occupancy in houses | ⊠ | ●●● |
| Redundancy and integration | Settlement or city development | Services integrated with housing | ●●● | ◎ |
| | | Infrastructure integrated with settlement | ⊠ | ◎ |
| | | Social amenities integrated in settlement | ●●● | ± |
| | | Livelihood proximity/considerations | ⊠⊠⊠ | ± |
| | | Redundancy integrated in housing or settlement | ◎ | ◎ |

| | | | | |
|---|---|---|---|---|
| Rights, equity, and security | Equity, inclusion | Recovery for all residents | ± | ± |
| | Dignity, sense of ownership | Participatory process | ●●● | ●●● |
| | Affordability | Low cost, affordable, and ease of maintenance | ◎ | ●●● |
| | Security | Better security of land, housing tenure | ± | ± |
| Resourcefulness (adaptive capacities) | Institutional capacities: recovery program, assistance, agility, rapidity, timely, yet strategic | Enforcement of planning and building regulations | ⊠⊠⊠ | ⊠⊠⊠ |
| | | Improved governance and institutional capacity | ◎ | ± |
| | | Proactive investment in prevention strategies | ◎ | ± |
| | Social capacities | Changed social mindset, norm, and improved or preserved social bond | ± | ◎ |
| | Individual capacities | Improved household awareness, security, and reduced aid dependency | ◎ | ◎ |
| | | Better quality of life and political voice | ◎ | ± |

decreased adherence to building standards. Redundancy was only incorporated at the house design scale in Bihar. Such outcomes highlight the importance of a settlement-based approach to integrate housing with necessary services, infrastructure, and nature-based solutions for disaster risk reduction.

*Rights, Equity, and Security*

The long-term rights and equity outcomes from housing recovery were mixed in both case studies, highlighting the centrality of land or tenure rights for resilient housing reconstruction. In both case studies, several socioeconomically weaker groups remained disadvantaged from not having received recovery assistance (e.g. in Bihar's Maanganj village), lack of timely recovery assistance (both case studies), or no land rights (e.g. in Gujarat's GIDC site). These outcomes can be attributed to the oversight by the state government in their recovery programming and the limited capacity of the local government in land management.

*Resourcefulness (Adaptive Capacities)*

In both case studies, in Gujarat and Bihar, the long-term capacities outcomes from housing recovery among government authorities, CSOs, the construction sector, and communities show improvement, yet are mixed. In Gujarat since the 2001 earthquake, institutional capacities and CSO capacities have improved over time with GSDMA continuing to share lessons learned with other states and CSOs such as Abihyaan being invited by other state governments (e.g. Bihar) to offer advice on recovery policy formulation. However, in Bihar, it is challenging to suggest any such improvements in institutional capacities. Although capacities within the construction sector and communities have improved, in both case studies, they are slipping back to pre-disaster construction standards. Only in sporadic instances there is an ongoing presence and sustained capacity building of communities by CSOs with/without the support of state government. Such capacity-building activities include linking skills training with educational institutions, providing funding, or other efforts extending beyond the timeframe of housing reconstruction.

### *Key Factors with the Potential to Catalyze Resilience Outcomes in the Long Term*

#### *Leading Role of Government, a Dedicated Authority for Disaster Management, and Adoption of an Owner-Driven Housing Reconstruction Program*

In 2001, India took a giant leap in their housing reconstruction program and disaster management in three aspects: (i) establishing a lead authority for disaster management; (ii) creating partnerships with the private sector and CSOs; and (iii) adopting an owner-driven housing reconstruction (ODHR) approach at a large scale.

Having a dedicated authority for disaster recovery management in both Gujarat and Bihar, the Gujarat State Disaster Management Authority (GSDMA) and Bihar State Disaster Management Authority (BSDMA), respectively, was vital for long-term success of the recovery programs.

These dedicated state government authorities, that exist to date, have fostered a culture of continuous learning and capacity building within government. This is evident in improvements made in ODHR policy from 2001 to 2008, by incorporating learnings from the Gujarat state government and CSOs in 2001, carried out by the Bihar state government in 2008. While an ODHR approach was adopted in both states and reconstruction projects were community participatory in nature, incorporating skills training and capacity building, ODHR in Bihar was marked by the agility of piloting prior to policy formation and decentralized governance (e.g. *Kosi Setu Kendras)*, which can be understood as being a highly sophisticated approach, not yet evident in post-disaster recoveries. Changes were made to the financial package in Bihar to include daily wages and land tenure security, especially for low-income and landless groups. Overall, the development of ODHR from 2001 to 2008 represented advancement in recovery approaches towards a post-development agenda. Although there were positive aspects of such agility in the ODHR program, such as the development of a more contextually appropriate and resilient rebuilding program, this agility also had downsides such as the delay in recovery of the vast majority of flood-affected residents in Bihar, who did not receive assistance for nearly two years after floods.

*Partnership-Based Governance and Coordination*
Governance and coordination played a crucial role in navigating the complexity of disaster reconstruction (including funding from multiple sources and managing a multitude of CSOs and donors), with both Indian case studies demonstrating their successes and failures.

In 2001, post Gujarat Earthquake, a top-down governance and coordination mechanism was set up, with clear roles for all tiers of the government and CSOs. This hierarchical governance structure ensured that the state government oversaw policy formulation and distribution of financial assistance to survivors and the CSOs adopted villages or urban settlements to assist earthquake-affected communities with socio-technical and additional financial support to rebuild their housing. However, various implementers (CSOs and private sector), with diverse funding sources, resulted in a wide variety of practices for residents to choose from, as discussed by Vahanvati and Mulligan (2017). For example, in Gujarat, some of the options residents had to choose from included provision of a fully built house, to materials provision along with construction supervision or technical skills training. Contrary to that, in 2008 post Kosi River floods, a decentralized governance and coordination mechanism was set up, whereby the state government and CSOs worked as partners in developing recovery policy, distributing uniform financial assistance to all affected residents based on house damage assessment, and enabling communities with decision making for rebuilding a safer house.

The recovery coordination efforts in both Indian case studies were jointly done by the state and local authorities with the support of CSOs. However, from 2001 to 2008, coordination took a leap from Bhuj's informal *Setu Kendras* set up at a local regional level, to Bihar's scaled-up, institutionalized and formalized *Kosi Setu Kendras*. In 2001 in Kachchh, Abhiyan had informally created *Setu Kendras* (meaning bridging hubs), for coordinating the massive task of recovery and managed an influx of international and local CSOs, potentially also due to a lack of such coordination by the local government. *Setu Kendras* were highly successful in acting as a conduit between communities and government, helping both with smoother information sharing, communication, grievances, or needs resolution, as well as playing the vital role of one-stop shops providing communities with vital recovery support. Underscored by the success of *Setu Kendras* in Bhuj, in 2008 in Bihar *Kosi Setu Kendras* (KSKs) were established at scale throughout flood-affected districts, and were institutionalized, that is managed and funded by the Bihar state government. The KSKs worked effectively during Phase I, under the supervision of CSOs, who were the ones with extended experience in reconstruction implementation. However, in Phase II, under the management of the state government, these KSKs lost community trust as they became infiltrated with corruption. Having said that, a decentralized governance and a shelter hub led coordinated recovery in Bihar led to uniform implementation of housing reconstruction and assistance practice, as discussed by Vahanvati and Mulligan (2017). While there is merit in decentralized coordination mechanisms such as KSKs in Bihar, concerns emerge about their effectiveness when such hubs are institutionalized and fully embedded in government (especially government bodies that lacked recovery coordination or implementation capacity), at risk of corruption and social power dynamics leading to unequal resource distribution and the entrenching of vulnerability.

There was continuity from 2001 to 2008 in terms of the government's decision to borrow money for recovery from the World Bank and make it readily available to affected residents directly, via their bank accounts. Moreover, to ensure gender equity, bank accounts were mandated to be in the names of both men and women spouses of the family. While the transfer of money directly from government into residents' banks allowed for maintaining transparency, it also led to further liberalizing of the economy. The dispensation of financial aid in tranches, contingent upon the completion of reconstruction milestones with compliance verification from government engineers, encouraged adherence to robust building practices, and hence, is worth replicating during future recovery efforts. However, both cases exposed disparity in financial package provision, whether it was homeowner versus renter or lower socioeconomic group based bias in Gujarat or social power-based bias in Bihar due to the prevalence of corruption. It also resulted in eroding public trust in the government and has diminished the government's efficacy in program evaluation and disaster risk reduction strategy formulation. Such power dynamics, entangled with financial aid-related corruption, is a major issue in developing countries like India. India could benefit from efforts put towards equitable and transparent governance structures as they play a vital role in achieving resilience outcomes.

### *Active Role of CSOs in Voicing Community Concerns and Capacity Building*

India has a strong presence of civil society organizations (CSOs) based on the concepts of *daana* (giving) and *seva* (service). CSOs played an active role in shaping housing reconstruction policies in Gujarat (through negotiations) and in Bihar (in collaboration), creating an enabling environment for disaster-affected communities and building capacities of government authorities as well as construction labor. Hence, understanding the role of CSOs becomes important.

From 2001 to 2008, the role of CSOs advanced from being mere implementers in Gujarat to policy advocates and government's capacity enablers in Bihar. In 2001, CSOs in Gujarat amplified community voices and concerns, leading to the state government of Gujarat changing their housing reconstruction policy to become owner-driven. In contrast, in 2008 due to their immense experience and knowledge of best-practice owner-driven reconstruction, CSOs were invited to advise Bihar's state government and build their capacity.

CSOs were integral, in both cases, in creating an enabling environment for communities, and building capacities of builders or construction laborers for rebuilding safer housing. Since the majority of houses are built by people through "people processes" or by the informal construction sector, robustness in housing is a major concern. Hence, CSOs have been actively providing handholding support to communities as well as skills training to construction laborers, integrating local traditional knowledge with an understanding of hazard safety features (see detailed discussion in sustained capacity building, later). CSOs also advocated for the rights of marginalized groups, since the Indian society has stratified communities based on class, religion, and gender. In both cases, CSOs have ensured land rights of socioeconomically weaker groups, financial assistance for the landless, accountability in recovery programs, and implemented community-centered, inclusive recovery, which all merit continuation.

### *Land Use and Settlement Planning*

Land use planning at city scale and master planning of settlements is essential for reducing hazard exposure, building in redundancies for hazard mitigation (e.g. mangroves for flood mitigation), as well as integrating infrastructure and services.

In India, land use planning changes that can trigger settlement layout changes or relocation for hazard exposure reduction has not been practiced widely. Out of the two case studies, such practice is only evident in Bhuj, post 2001 Gujarat Earthquake. In Bhuj's walled city, land readjustment and urban renewal was undertaken by the Bhuj Area Development Authority (BHADA) with help of the private sector for multi-hazard (earthquake, drought, and high wind) risk reduction. Settlement layout was done for the GIDC relocation site (a greenfield site) by a CSO—Abhiyan, who created comprehensive settlement plans, integrating open green spaces, social infrastructure, services, and infrastructure. While social infrastructure is known to improve social networking and social capital, other infrastructure such as roads, internet, and transport options determines people's movement, essential for livelihood. Moreover, COVID-19 has highlighted the importance of proximity to open green spaces for societal health and well-being. Internationally, a settlement-based or area-based approach is increasingly promoted (Urban Settlements Working Group, 2020), particularly in developing countries, due to an understanding of interdependence between physical infrastructure and social dynamics.

*Centrality of Land or Tenure Rights and Gender-Based Rights for Housing Resilience*

Secure land tenure was identified as the bedrock of resilience in both case studies as it provides long-term stability and a sense of ownership among residents. In 2001, the rights of renters, those living in informal settlements, and lower socioeconomic groups, were not considered in the ODHR recovery program until CSOs pressured the government to include them. To date, challenges persist in the GIDC relocation site in Gujarat, with residents selling their properties informally, or renting them out, due to inadequate land titles, and thus house ownership issues.

Based on the learnings of CSOs and the government from Gujarat's recovery in 2001, issues of land rights were addressed in Bihar post 2008 floods, to ensure the inclusion and rights of marginalized communities. In Bihar, flood-affected communities were enabled to be inclusive by resolving land rights among themselves for housing reconstruction to begin. Moreover, unlike in Gujarat, where the renters or landless were forgotten in the recovery assistance package, in Bihar the recovery policy included additional financial assistance for the landless from the government from the very start. The long-term outcomes of secure land rights have created a sense of security for residents in Bihar's Maanganj village; however, that is not the case for residents living in Gujarat's GIDC site. These two case studies demonstrate that mere rebuilding of robust housing or structures capable of withstanding natural hazards isn't enough, because evidence proves that true resilience is built upon the realization of residents' fundamental rights, which underlines their long-term stability and security.

*Understanding Social Systems as Integral to the Success of Housing Reconstruction Projects*

In both case studies, social issues emerged as pivotal in resilient recovery in India. India's social caste-based hierarchy influenced disparity in recovery assistance distribution, particularly in Gujarat post 2001 Earthquake, while corruption and power dynamics created similar disparity in Bihar post 2008 floods. The social fabric of the community was strained in Gujarat's GIDC relocation site due to the mixed caste-based communities, while it was strengthened in the wake of recovery efforts in Bihar. Therefore, in the Indian context, having a deeper understanding of social bonds and cohesion as cultural aspects is integral to a holistic recovery planning that has the potential to succeed in regard to community resilience over a longer period.

*Sustained Capacity Building, Particularly For a Strong Culture of Compliance with Building Standards, Supported by the Government*

Sustained education, awareness raising, skills training, monitoring, and capacity building within the construction sector have emerged as essential for maintaining and advancing housing resilience practices, in a society that is predominantly informal and lacks compliance with building standards. Both case studies illustrate the achievement of robust housing structures. However, in Gujarat post 2001 earthquake, capacity-building activities for construction labor were sustained over a longer period by a few CSOs. As noted by Vahanvati and Mulligan (2017), in Gujarat the government partnered with vocational institutions, allowing CSOs to help trained masons gain seismic-safe construction certification at international standards. Due to certification, trained masons from Gujarat have been spreading knowledge of quality and safe construction skills among emerging masons informally. At the GIDC settlement, while the CSO did not link trained residents with education for certification for their livelihood, continued presence of the local CSO meant that people could knock on their doors for assistance, over a longer period after reconstruction was complete. Sustained capacity building has the potential for continuous transformation within housing systems, needed for futureproofing communities in the face of climate change.

Among communities, a lack of continuous awareness-raising activities or capacity-building activities are identified as barriers to residents' ability to make risk-informed decisions, as well as learn new skills to maintain newly introduced green technologies. In the short term, there was an overall increase in residents' awareness of hazard safety features in housing due to active participation in the decision-making process throughout their own housing reconstruction process. However, in the longer term, such awareness or maintenance capacity is fading. For example, green construction technologies introduced in Bihar after the 2008 Kosi River floods, such as bamboo construction, waterless toilets, or solar lighting, face major maintenance challenges, causing most of them to be discontinued or left in a state of disrepair. Similarly, in Gujarat's GIDC site, post 2001 earthquake, the green sewage water recycling system and garbage conversion to vermicompost have also become dilapidated due to maintenance challenges.

At the government level, having a comprehensive disaster management framework and a dedicated recovery authority is helpful in continuous capacity building for improvement in policies and practices based on learnings from past disasters (as discussed in earlier section). However, some of the key barriers to stronger institutional capacities for resilience building are a lack of reliable hazard risk mapping to make risk-informed decisions about land use planning changes, a lack of a culture of strict enforcement of building codes or standards, and limited support for sustained capacity building of local government authorities.

## Conclusion

The authors aimed to identify key factors that helped or hindered systemic changes, through post-disaster housing reconstruction projects, from a rights and resilience perspective in the long term (at least 10 years since the disasters) in India. Valuable lessons emerged from an examination of long-term outcomes of two housing reconstruction project case studies after both disasters in India—the 2001 Gujarat Earthquake and the 2008 Kosi River floods in Bihar (see Fig. 2.1). Looking back at the housing reconstruction policies and projects in the urban Bhuj and rural Maanganj, following disasters that occurred almost 24 and 17 years ago, respectively, key factors with potential to catalyze systemic change are identified.

The first key factor includes having a comprehensive disaster management framework and a dedicated government authority to lead recovery efforts, which India has established and continues to date, fostering a culture of continuous learning from past interventions for resilience building. A lead role of the government has led India to adopt an owner-driven housing reconstruction program that is participatory, inclusive, and focused on strengthening residents' capacities and capabilities (i.e. freedom to choose or make informed decisions). An owner-driven reconstruction program is essential in India since it creates space for CSOs in recovery practice and also because the majority of housing is built by the informal construction sector through a people's process, which requires training. The second key factor relates to the crucial aspect of recovery coordination and governance mechanism. All Indian case studies have highlighted an innovative one-stop shop or shelter hub, with ideal composition of helpers enabling easy communication between government and communities, leading to smoother and safer housing reconstruction. Indian case studies encapsulate the importance of government's recovery programs incorporating agility, inclusivity, and collaborative governance (e.g. in Bihar), as they are critical drivers in the holistic model for recovery that empowers communities and safeguards the environment.

The third key factor is the active role of CSOs in India, as they amplify the voices of the often forgotten, lower socioeconomic groups, to ensure inclusive and equitable recovery for all. India's CSOs have vast experience in implementing ODHR by creating an enabling environment and capacity strengthening of communities as well as government officials. The fourth key factor is the need to improve land use planning at the city scale and master planning of settlements to reduce hazard exposure and integration of basic services and infrastructure, essential for residents' daily life and livelihood. However, India lags in land use planning changes as well as building standards enforcement, both of which are essential for reducing hazard exposure and robustness of housing.

The fifth and sixth factors demonstrate land rights protection and social capital enrichment as essential pillars on which robust housing and resilience outcomes can be built. Equitable resource distribution and inclusive practices can together support a community's capacity to withstand and adapt to future adversities, while their absence can exacerbate social inequity and vulnerability.

The seventh and last key factor highlights the significance of sustained learning for capacity building to convert to capacity development. However, such efforts are occurring sporadically in India. The Indian case studies highlight a need for hazard risk mappings for informed land use planning, hazard preparedness including early warning systems, community awareness, and skills training in the construction sector, combined with improvement in the enforcement of building standards in new developments. All of these are essential for housing resilience, but they rely on continuous capacity building and embedding of those skills in livelihoods for capacity development. Insights distilled from Bhuj City's and Maanganj village's recovery experiences highlight that a reimagined approach to housing reconstruction, along with a dynamic practice of continuous learning and improvement, can lay a foundation for a more equitable, sustainable, and resilient society.

## References


Arya, A., Boen, T., & Ishiyama, Y. (2014). *Guidelines for Earthquake Resistant Non-Engineered Construction*. Retrieved from https://reliefweb.int/report/world/guidelines-earthquake-resistant-non-engineered-construction

Balachandran, B. R. (2005). *The Reconstruction of Bhuj: Integration of Disaster Mitigation into Planning and Financing Urban Infrastructure after an Earthquake* (pp. 95–124). Retrieved from https://www.urban-response.org/system/files/content/resource/files/main/944.pdf

Barenstein, J. D. (2006). Housing Reconstruction in Post-earthquake Gujarat: A Comparative Analysis. *Network Paper, 54*. Retrieved May 2019, from https://odihpn.org/wp-content/uploads/2006/04/networkpaper054.pdf

BHADA. (2005). *Earthquake 2001*. Bhuj Area Development Authority. Retrieved from https://bhujada.com/earthquake-2001/

BHADA. (2016a). *About BHADA*. Bhuj Area Development Authority. Retrieved from https://bhujada.com/about-bhada/

BHADA. (2016b). *Relocation Sites*. Bhuj Area Development Authority. Retrieved from https://bhujada.com/relocation-sites-2/

BSDMA. (2012). *Evolution of Bihar State Disaster Management Authority (BSDMA)*. Government of Bihar. Retrieved August 10, 2018, from http://bsdma.org/AboutUs.aspx?id=2

CAPAM. (2004). *CAPAM International Innovations Awards Programme 1998–2004*. Retrieved from www.capam.org. www.unpan.org/innovmed/documents/awards.pdf

Census of India. (2011). *Villages & Towns in Tribeniganj Block Supaul, Bihar*. Retrieved from https://www.censusindia.co.in/villagestowns/tribeniganj-block-supaul-bihar-1111

EPC, & TCG. (2004). *Participatory Planning Guide for Post-Disaster Reconstruction*. Environmental Planning Collaborative & TCG International.

Flood Hazard Atlas—Bihar: A Geospatial Approach, 1–169. (2020). Retrieved from https://reliefweb.int/report/india/flood-hazard-atlas-bihar-geospatial-approach; https://reliefweb.int/updates?list=India%3A%20Floods%20-%20Jun%202008%20Updates&advanced-search=%28D5430%29

FMIS, & GoB. (2009). *Flood Report 2009.* Retrieved from http://fmis.bih.nic.in/Report2009.pdf

GoB, & ODRC. (2008a). *Reconstruction and Rehabilitation of Kosi Flood Affected Regions in Kosi Division: Memorandum of Understanding (MOU).* G. o. Bihar.

GoB, & ODRC. (2008b). *Workshop on Owner Driven Reconstruction and Rehabilitation for Kosi Flood Affected Regions* [Technical Guidelines].

GoB, & World Bank. (2010). *Bihar Kosi Flood Recovery Project.* Government of Bihar.

GoB, World Bank, & GFDRR. (2010). *Bihar Kosi Flood (2008) Needs Assessment Report.* World Bank. Retrieved from https://www.gfdrr.org/sites/gfdrr/files/publication/GFDRR_India_PDNA_2010_EN.pdf

GoI, & UNDP. (2011). *Disaster Management in India.* Nation Centre for Disaster Management. Retrieved from http://nidm.gov.in/books.asp

GSDMA. (2001a). *About Gujarat State Disaster Management Authority (GSDMA).* Government of Gujarat. Retrieved March 10, 2014, from www.gsdma.org/about-us/introduction.aspx

GSDMA. (2001b). *GEERP and GERRP.* Government of Gujarat. Retrieved March 10, 2014, from http://gsdma.org/Content/geerp-gerrp-4216

Kumar Bharti, N., Chancel, L., Piketty, T., & Somanchi, A. (2024). Income and Wealth Inequality in India, 1922–2023: The Rise of the Billionaire Raj (Working Paper, Issue 09). The World Inequality Database (WID). https://wid.world/newsarticle/inequality-in-india-the-billionaire-raj-is-now-more-unequal-than-the-british-colonial-raj/

Meena, M., Singh, S., & Bhattacharyya, R. (2014). After the Kosi Avulsion, August 2008, North Bihar, India: Notes on the Socio-Economic Impact. *Asian Profile, 42*(3). Retrieved from https://papers.ssrn.com/sol3/papers.cfm?abstract_id=2764125

Mucke, P., Radtke, K., Kirch, L., Hamilius, N., Jeschonnek, L., Gruden, N., Atwii, F., Sandvik, K. B., Kirch, L., Paragi, B., Radtke, K., Schneider, S., & Weller, D. (2022). *The World Risk Report* (978-3-946785-14-9). B. E. Hilft. Retrieved from https://www.preventionweb.net/publication/worldriskreport-2022-focus-digitalization#:~:text=Some%20key%20findings%20of%20the,populous%20countries%20in%20the%20world

Mukherji, A. (2008). *Negotiating Housing Recovery: Why Some Communities Recovered While Others Struggled to Rebuild in Post-earthquake Urban Kutch, India University of California, Berkeley.* UMI dissertation publishing by ProQuest.

NDMA. (2005). The Disaster Management Act 2005. New Delhi, India: National Disaster Management Authority, Government of India. Retrieved from https://ndma.gov.in/Reference_Material/DMAct2005

NIDM. (2010). *National Disaster Management Guidelines: Role of NGOs in Disaster Management*. G. o. I. Nation Disaster Management Authority. Retrieved from http://nidm.gov.in/PDF/guidelines/NGO_Guidelines.pdf

ODR Collaborative. (2008). *Proceedings: Eastern region Consultation on ODR*. ODR Collaborative. Retrieved from http://odrcollaborative.net/odrc_docs/ODRkolkataSummary.pdf

ODR Collaborative. (2010). *Guidelines for House Construction using Bamboo (in Hindi)* [Technical Guidelines]. Department of Planning and Development, Government of Bihar.

Price, G., & Bhatt, M. (2009). *The Role of the Affected State in Humanitarian Action: A Case Study on India*. Overseas Development Institute (ODI)/Humanitarian Policy Group (HPG). Retrieved from http://dspace.cigilibrary.org/jspui/bitstream/123456789/24119/1/Role%20of%20the%20affected%20state%20in%20humanitarian%20action.pdf?1

Pritchard, B., & Thielemans, R. (2014). 'Rising Waters Don't Lift All Boats': A Sustainable Livelihood Analysis of Recursive Cycles of Vulnerability and Maladaptation to Flood Risk in Rural Bihar, India. *Australian Geographer, 45*(3), 325–339. https://doi.org/10.1080/00049182.2014.930001

Shah, K. (2009). *Architects and Architectural Practice in India: Some Imperatives* (Community Architects Report 2010–2012) (not 100% sure of the year), Issue. ACHR. Retrieved from https://architexturez.net/doc/az-cf-166148

Suri, S. (2023). *The Implications of the Growing Population on Human Development in India*. Observer Research Foundation. Retrieved February 15, 2024, from https://www.orfonline.org/expert-speak/the-implications-of-the-growing-population-on-human-development-in-india

Taheri Tafti, M., & Tomlinson, R. (2016). Long-Term Socio-Spatial Transformation of Earthquake-Affected Neighbourhoods in Bhuj: Who Stayed, Who Left and Who Moved In? *International Journal of Disaster Resilience in the Built Environment, 7*(3), 230–258. https://doi.org/10.1108/IJDRBE-07-2014-0049

UN-DESA. (2023). *Policy Brief No. 153: India Overtakes China as the World's Most Populous Country*. Retrieved from https://www.un.org/development/desa/dpad/wp-content/uploads/sites/45/PB153.pdf

UNDP. (2001). *From Relief to Recovery: The Gujarat Experience*. UNDP. Retrieved from https://www.undp.org/india/publications/relief-recovery-gujarat-experience

UNDP. (2009). *Kosi Floods 2008: How We Coped! What We Need? Perception Survey on Impact and Recovery Strategies*. UNDP. Retrieved from www.in.undp.org/content/dam/india/docs/kosi_floods_2008.pdf

UNDP. (2014). *About Bihar*. UNDP India. Retrieved October 15, 2014, from www.in.undp.org/content/india/en/home/operations/about_undp/undp-in-bihar/about-bihar/

UNDP. (2023). *India Ranks 132 on the Human Development Index as Global Development Stalls*. Retrieved November 2023, from https://www.undp.org/india/press-releases/india-ranks-132-human-development-index-global-development-stalls#:~:text=India's%20HDI%20value%20of%200.633,a%20decent%20standard%20of%20living

UNDP. (2024). *Climate Promise—India*. Retrieved January 8, 2024, from https://climatepromise.undp.org/what-we-do/where-we-work/india

Unnati. (2008). In M. International (Ed.), *Disaster Risk Reduction: A compilation of literature for use of practitioners*. Unnati—Organisation for Development Education.

Urban Settlements Working Group. (2020). *Settlement Approaches Guidance Note*. Global Shelter Cluster. Retrieved from https://sheltercluster.org/settlements-approaches-urban-areas-working-group/documents/settlement-approach-guidance-note

Vahanvati, M. (2015, 17–21 September). *Reviving People's Trust in Bamboo Technology: A Case-Study of Orlaha Settlement Reconstruction in Bihar, after the 2008 Kosi Floods*. In 2015 World Bamboo Congress, Damyang, South Korea.

Vahanvati, M. (2018a). A Novel Framework for Owner Driven Reconstruction Projects to Enhance Disaster Resilience in the Long Term. *Disaster Prevention and Management, 27*(4), 421–446. https://doi.org/10.1108/DPM-11-2017-0285

Vahanvati, M. (2018b). *Thesis: Owner-Driven Housing Reconstruction as a Means of Enhancing Disaster Resilience of At-risk Communities in India* [RMIT University]. Melbourne, Australia. Retrieved from https://researchrepository.rmit.edu.au/discovery/fulldisplay/alma9921864052701341/61RMIT_INST:ResearchRepository?tags=scholar

Vahanvati, M., & Beza, B. (2017). An Owner-Driven Reconstruction in Bihar [Scholarly]. *International Journal of Disaster Resilience in the Built Environment, 8*(3), 306–319. https://doi.org/10.1108/IJDRBE-10-2015-0051

Vahanvati, M., & Mulligan, M. (2017). A New Model for Effective Post-disaster Housing Reconstruction: Lessons from Gujarat and Bihar in India. *International Journal of Project Management, 35*(5), 802–817. https://doi.org/10.1016/j.ijproman.2017.02.002

WMO. (2021). *State of the Climate in Asia 2020*. World Meteorological Organization. Retrieved from https://library.wmo.int/viewer/57695?medianame=1273_State_of_the_Climate_in_Asia_2020_en_#page=2&viewer=picture&o=bookmarks&n=0&q=

World Bank. (2018). *India: Bihar Kosi Flood Recovery Project*. Implementation Completion and Results Report (ICR) Issue. W. Bank. Retrieved from http://documents.worldbank.org/curated/en/716681545409586198/India-Bihar-Kosi-Flood-Recovery-Project

World Bank. (2022). *Population Living in Slums (% of Urban Population)*. The World Bank Group. Retrieved from https://data.worldbank.org/indicator/EN.POP.SLUM.UR.ZS

World Bank. (2024). *Population Density (People per sq. km of Land Area)*. World Bank. Retrieved from http://data.worldbank.org/indicator/EN.POP.DNST

World Bank, & ADB. (2001). *Gujarat Earthquake Recovery Program* [Electronic]. Retrieved from www.preventionweb.net/files/2608_fullreport.pdf

World Bank, & GFDRR. (2013). *Enhancing Safety and Improving Living Conditions for 100,000 Vulnerable Families: Building a Culture of Resilient Reconstruction in India*. Retrieved from https://www.gfdrr.org/sites/default/files/publication/Project_Highlights_Owner-Driven_Housing_Reconstruction_0.pdf

CHAPTER 3

# Long Term Post-Disaster Housing Reconstruction in Thailand: Case Studies (Upper-Middle-Income Country)

**Abstract** In the context of Thailand's emerging disaster management and recovery systems, this chapter looks back at two cases of long-term housing reconstruction, in southern Thailand after the 2004 Indian Ocean Tsunami and in Central Thailand after the 2011 floods (Fig. 3.1), to clarify key lessons and factors that enabled resilient outcomes. The first case study area was selected as Namkhem Village in Phang Nga Province after the tsunami, where 187 residents and stakeholders were interviewed between 2013 and 2022. The second case study was Ayutthaya Province after the 2011 flood, with a focus on the Huntra sub-district. Eighty residents and stakeholders were interviewed in 2011, and secondary data was collected from village number 1 (779 households) regarding flood relief and financial support provided by the government, and 40 residents were interviewed again in 2023. Examining roles of stakeholders, this chapter considers changes in processes and policies for reconstruction toward future disaster prevention and reflects on their effectiveness in supporting communities after disaster. A comparison between the two cases in Thailand highlights several key factors that contribute to resilient long-term outcomes. Local government and community leaders were crucial in driving housing reconstruction, overseeing construction, fostering strong community ties, and ensuring recovery efforts were sustainable and aligned with residents' needs. Donor support also played a critical role in tsunami recovery, where professional construction and ongoing support



M. Vahanvati et al., *Resilient Recovery from Disasters*,
https://doi.org/10.1007/978-981-97-8047-1_3

for housing and infrastructure improvements helped create a more robust and adaptable community. Overall, housing recovery efforts in Thailand demonstrate the importance of local leadership, community engagement, and long-term planning for achieving resilient outcomes. They also show the varied support required for communities with varied socio-economic status and corresponding disaster vulnerabilities. While challenges remain, particularly in terms of equity and the consistent application of risk management principles, experiences from these disasters provide valuable lessons for future recovery efforts.

**Keywords** Thailand • Community participatory • Post disaster • Housing reconstruction • Disaster resilience • Long-term recovery

## Thailand's Hazard and Disaster Management Context

With a relatively recent history of major hazard events, the first recorded mega-disaster in Thailand was the 1962 devastating tropical storm surge Harriet, which struck Laem Talum Puk peninsula in Nakorn Sri Thammarat Province in the south, resulting in the loss of 769 lives. Since then,

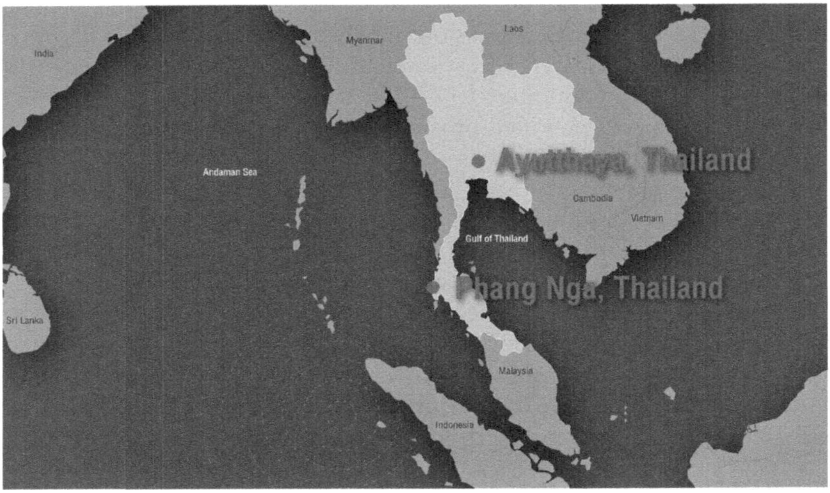


**Fig. 3.1** Map of Thai case studies (Source: Authors)

Thailand has been affected by various natural hazards, including floods, droughts, storms, landslides, and tsunamis.

Early legislation (1939–1979) related to disaster management included the Air Attack Prevention Acts enacted in 1939 and 1944 during World War II, the Local Patrol Regulation introduced in 1954 for fire response, and the Civil Threat Prevention Act enacted in 1979, which marked the first comprehensive law for managing disasters in Thailand, initiated by the Ministry of Interior. More recently, National Accident Prevention policies were updated in 1983, 1995, and 2002, reflecting ongoing efforts to improve national safety and disaster preparedness.

Thailand's disaster management policies have often been reactive rather than proactive, with significant legislative changes frequently occurring after major disasters. For example, the 2002 establishment of the Department of Disaster Prevention and Mitigation (DDPM), to centralize disaster management efforts and improve coordination, and subsequent reforms post-2004 tsunami represent critical turning points in improving the country's disaster preparedness and response capabilities. Continuous updates and additions to disaster management plans, particularly after the 2011 floods, highlight the government's efforts to address evolving challenges and enhance resilience against future disasters.

## Thailand's Disaster Management, Response Capabilities, and Housing Context

New government regulations implemented in response to the 2004 tsunami reflected a shift towards centralized command in disaster response. A 2005 regulation designated the Prime Minister's Office to manage national disasters and the 2007 Disaster Prevention and Mitigation Act covered comprehensive disaster preparedness and response strategies. Showing a proactive approach toward future tsunami risk reduction, the Tsunami Master Plan (2009–2013) was developed as a dedicated plan for tsunami mitigation. Subsequently, the National Disaster Prevention and Mitigation Plan (2010–2014) provided a structured framework for disaster management across various levels of government.

Further revisions in disaster management policies after the 2011 flood disaster emphasized effective water management and flood disaster response. In 2013, A Policy Update: A new chapter on "Water Management and Flood Disaster" was added to the 2010–2014 Plan, indicating continuous improvement and adaptation of policies in response to recent disasters. The Master Plan of Tsunami Disaster Prevention (2009–2013) is Thailand's comprehensive disaster management policy, providing a structured approach to mitigating tsunami impacts through prevention, preparedness, emergency management, and recovery phases. This plan highlights Thailand's commitment to improving disaster management and protecting coastal communities from future tsunami threats, and the inclusion of detailed risk maps and involvement of multiple collaborating agencies demonstrate a thorough and coordinated effort to enhance tsunami resilience in Phang Nga Province.

However, there are flood management gaps in Thailand's National Disaster Prevention and Mitigation Plan 2010–2014. While the plan outlines critical components such as water management and warning systems and addresses flood disaster management at various levels, it falls short in the following key areas. (1) It lacks detailed local data, such as granular flood information at the local scale, crucial for effective management and response. (2) There is no record of water depth in each affected area, a critical parameter for understanding the severity and impact of floods. (3) It inadequately addresses the recovery phase, leaving gaps in post-disaster management and long-term resilience-building. (4) Compared to other disaster plans, such as for tsunamis, the flood management plan lacks comprehensive flood-risk data, essential for effective risk assessment and mitigation.

Disasters in Thailand are categorized into four levels: level 1 can be managed by the local community, level 2 is managed by the district government, level 3 is managed at the provincial level, and level 4 is a national-level disaster requiring national government management. Thailand's disaster management structure includes specific roles for each government level, from national to local government. At the top or national level (level 4), the prime minister directly commands the national response for large disasters under a single command system. At the next level (level 3), the Ministry of Interior acts as a central support system and intermediary for the national government's directives, and the DDPM plays a central role in executing disaster prevention and mitigation strategies. At the provincial level (level 2), disasters are managed by the provincial Disaster Management Committee, which reports to the Ministry of Interior, along with the provincial Mayor. For example, in Bangkok, the Bangkok Disaster Committee and the Bangkok Mayor managed disaster response. At level 1, the district Mayor manages disaster response within their district, and local government officials, including city mayors, chief sub-districts, and the Bangkok City mayor, handle local disaster management.

Post-disaster housing reconstruction projects in Thailand are varied, and residents experience some flexibility in their choices. Based on residents' requests and negotiations with the government, housing reconstruction options include provided houses, relocated houses, and self-built houses. Although there is some research on recovery in Thailand after the 2004 tsunami and 2011 flood, those studies have predominantly focused on damages and effects on livelihood, with limited focus on housing reconstruction and residents' living conditions.

## Case 1: Housing Reconstruction After the 2004 Indian Ocean Tsunami

Resulting from the massive undersea "Sumatra Earthquake" the Indian Ocean Tsunami struck on December 26, 2004, causing widespread devastation to coastal areas around the Indian Ocean, obliterating buildings, infrastructure, and entire communities. The tsunami killed 230,000 people and affected 14 countries in the Indian Ocean. For Thailand, this was the first recorded tsunami and was also the most destructive disaster in terms of human losses, killing 8345 people in six provinces of southern Thailand on the Andaman coast, including visitors from around the world who had traveled to Thailand during the peak tourist season (Fig. 3.2).

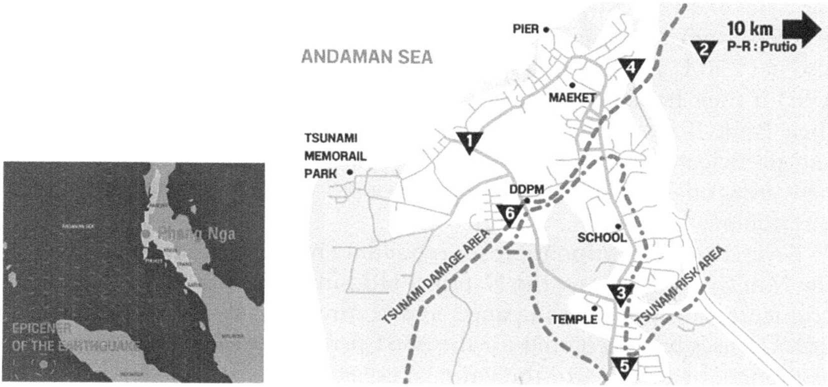


Fig. 3.2 Sumatra Earthquake, Indian Ocean Tsunami 2004 (left), and map of post-disaster housing projects in Namkhem Area (right) (Source: T. Sararit)

### *Emergency Management after the Unprecedented 2004 Indian Ocean Tsunami Disaster*

After the tsunami, the national government promptly activated the existing "civil defence emergency" system; the prime minister directly commanded six key ministries—Ministry of Interior, the Ministry of Health, the Ministry of Foreign Affairs, the Department of Disaster Prevention and Mitigation (DDPM), the Thai Military Force, and the Ministry of Social Development and Human Security—to establish a tsunami committee. The immediate priority was to search, rescue, and identify persons, and provide shelter for the displaced. The UN and various international organizations also supported the response and recovery phases, including housing assistance, supporting children, and economic recovery; many local and international non-government organizations (NGOs) also contributed donations to assist tsunami-affected residents. The Ministry of Social Development and Human Security was tasked with coordinating the efforts of various government agencies to provide shelter to those displaced by the tsunami. Families whose houses had been destroyed or damaged were provided emergency shelter in relief camps, followed by temporary and permanent shelters.

### *Post-Tsunami Housing Reconstruction Support*

The government budget provided approximately 1.7 billion USD in funds for relief and reconstruction. Military forces and private companies oversaw repair or rebuilding of houses in each province. Each household could also get 750 USD in aid to repair their own house; or receive around 2500 USD if their house was completely destroyed to rebuild a new house on their land. If households had no land ownership, they could share land and participate in the construction of a "self-built" house, relocate to a new area, or move into a "provided house" built with funds from various donors.

Several agencies responsible for rebuilding houses included the military, the Ministry of Defence, the National Housing Authority, provincial governments, and private companies and contractors. By November 2005, 2688 houses had been built for the 3361 people who sought government assistance. In addition to the national government, the reconstruction of post-disaster permanent housing included many stakeholders, such as non-governmental organizations and residents' groups, among others.

### *Case Study: Housing Reconstruction in Namkhem Community, Phang Nga Province*

Namkhem, which means salt water, is the name of Village no. 2 of Bang Muang sub-district, in the Takua Pa district of Phang Nga Province, in south Thailand on the Andaman Sea. Namkhem is a valuable case study as a single village contains various types of housing recovery projects, including those where the government provided support in combination with support from villagers, NGOs, and the private sector. This study of the Namkhem community gathered long term recovery feedback from groups of residents in different housing relocation areas and projects, through interviews with 20 residents about their houses and life changes 20 years after the tsunami, which demonstrated that the funding source dictated the project requirements and outcomes.

#### *Pre-disaster Housing Context in Namkhem*

The Andaman fishing village of Namkhem was previously a mining industrial area from 1972 to 1982, attracting people from all provinces of Thailand to work in the mining industry. Later Namkhem developed into a fishing village along with a flourishing tourist business. Many workers came to Namkhem to work as laborers and lived in low-income communities or dense coastal slums before 2004.

#### *Disaster Impacts and Housing Recovery Policies and Programs*

The 2004 tsunami caused the death of 8345 people in Thailand, completely destroyed 3302 houses and damaged 1504 houses in Thailand. Phang Nga Province suffered the majority of housing damage, with 1904 houses destroyed and 604 houses partly damaged. Namkhem suffered the largest loss of life in the area including 824 casualties of people with legal registration and another 641 people without legal registration. In Namkhem, 1210 houses were completely destroyed and another 224 houses were partially damaged (Maitree, 2013, Namkhem community DDPM volunteer presentation).

Initially, survivors were accommodated in emergency shelters to ensure immediate safety and provision of necessities. Following the emergency phase, temporary houses were built to offer a more stable living environment while permanent solutions were being developed. Six months after the tsunami, significant and coordinated efforts by multiple government and non-government entities focused on transitioning residents from temporary housing to permanent homes. This process aimed to ensure stable and safe living conditions for all affected residents, whether they returned to their former land or relocated to new areas, and housing provision took several forms.

Starting six months after the tsunami, the first group of residents who preferred to return to their original locations moved back into new houses fully built by the army. One year after the tsunami, another group of residents was provided houses fully constructed by donor support in newly developed relocation areas. With the help of the Community Organizations Development Institute (CODI), tsunami survivors in a third group, excluded from the on-site or relocation housing reconstruction support schemes mentioned above, became part of a community-based process of self-built housing reconstruction.

*Post-disaster Reconstruction Project Implementation and Multiple Housing Options*

**Housing Provision On-Site**

For a significant number of residents with land ownership rights, government and donors from many non-governmental organizations offered some funding for "provided houses (non-relocated)" to be built on the residents' former land by the military, referred to as "military houses" (Sararit & Kondo, 2014). The Ministry of Interior provided the design of these houses and the army constructed about 800 of them (Fig. 3.3, left) in the Namkhem area and provided them to residents who did not

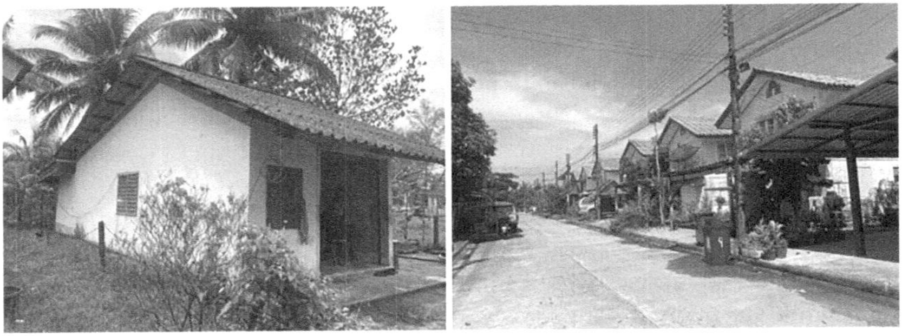

**Fig. 3.3** Provided houses (non-relocated) (left) and housing provided with land sharing (self-built relocated house) (right) in Namkhem. (Source: T. Sararit)

relocate. In the case of Phang Nga, the army was primarily in charge of rebuilding houses; houses built with army support have a "green stripe" of paint to show this.

### Housing Provision in Relocation Areas

In a second pattern, "provided-relocated" housing, residents were moved to a new location near Phru Tiao "Lake" on vacant land that belonged to the local government, with various donor groups providing funding for housing, and design and construction processes carried out by professional builders. Residents had no participation in decision-making, housing design, or construction of the houses, but as the project provided completed houses within one year, residents did not have to wait long to acquire permanent houses quickly. As the land belonged to the local government, this project also helped maintain the local population post tsunami. The local government also retained ownership of the houses; although officially, residents are not allowed to sell the houses to a third party, many sell them secretly, with agreements with new users that it is a rental house and that they are moving out to another place.

### Housing Provision with Shared Land

The low-income residents' group with no land ownership rights before the disaster had lived in the slums near Namkhem, working in the fishery industry. Without land ownership, they were ineligible to apply for post-tsunami housing provision. With support from CODI, it took them almost three years to complete the legal requirements and have a permanent place for housing reconstruction. Shared land was purchased using a community mortgage, and houses were built by the residents, with construction costs supported by donor funds. Residents paid the mortgage fees as a group each month; after 15 years, they gained ownership of the houses.

### Comparing Residents' Satisfaction Based on Housing Provision Type

For residents with official land ownership, most houses were rebuilt on former plots in about one year. Although this support was fast and residents received the houses relatively quickly, within six months to one year, the house construction quality was relatively low. However, it still took almost three years to complete the reconstruction of houses for displaced survivors because these residents had no land rights or land ownership and, in some cases, needed to relocate. Even though they did not need to pay for construction costs, residents of "provided housing", both in-situ and relocated, reported less satisfaction compared to those in self-built housing, across all three factors of building construction, economic and social relations, and disaster awareness. Without having a say in the house construction or opportunities to interact with neighbors, residents of "provided housing" felt disconnected from their own homes and neighboring communities.

However, for residents who participated in the housing construction processes, such as self-built housing projects, "relocation projects fostered stronger social relationships compared to "non-relocated" (in situ) housing recovery projects, as all residents started living in the new place together. Residents reported a stronger sense of ownership in the housing project, as they participated in every stage of the process and built the houses themselves.

*Long-Term Outcomes 20 Years Since the 2004 Tsunami*
Recovery in Namkhem included flexibility and was responsive to survivors' housing needs, with various housing options, including in situ and relocated houses, as well as provided houses and self-built houses with community-based processes. As a result of negotiations with the government, most residents could become housing beneficiaries, including renters, low-income residents, and fishermen living above the water, who had been initially excluded from the official recovery support plan. There was also customized support for indigenous communities.

Twenty years post tsunami, there were significant changes in the Namkhem community. In addition to developing various systems for evacuation drills, there have been improvements in infrastructure and public utilities in the area. Beyond these developments, a modern tsunami memorial has been established, part of education for the younger generation in the area. However, some areas such as those near the tsunami memorial and the Namkhem pier are underutilized. Furthermore, donated houses in the area have been abandoned, repurposed for storage, or rented out. Many communities in this area include low-income residents who cannot afford to relocate. New economic and tourism zones have emerged, such as the mangrove and Phru Tiao zones, which have become bustling residential areas with schools, public utilities, and road access to nearby tourist attractions. Some long-time residents have expanded or renovated their homes, and new tenants have moved into some of the residences. This contrasts with the first 10 years post tsunami, when many residents in relocation areas were dissatisfied because they were unfamiliar with their new neighbors. Additionally, the houses were not large enough or adequate for expansion.

Twenty years after the tsunami, relocation areas had become lively, with residents expanding and improving their homes, indicating a stronger intention to stay and a better relationship with neighbors. This outcome suggests that long-term planning is crucial in housing reconstruction projects. The changing economy, land use, and quality of life have led families to adapt more to their new environments, creating stronger community bonds. For future housing reconstruction, these findings suggest the importance of considering factors such as providing activities that allow disaster survivors to get to know each other when moving to a new area.

Another finding relates to to the quality of new houses in relation to the timing of their construction. Detached houses built a little later after the tsunami tended to be of higher construction quality compared to houses quickly provided by the government initially. As they are more durable, residents take more pride in their homes. Although relocated houses were initially identical, as homeowners' incomes increased, they renovated, extended, and decorated their homes, demonstrating their satisfaction and ability to improve and maintain their own living spaces.

Long-term outcomes also demonstrated the importance of adaptation in the design of housing plots for post-disaster housing reconstruction. For example, allowing future extensions could greatly support the development of residents' housing environment, increase satisfaction levels, and encourage better community participation.

**Robustness and Diversity (Buffer Capacities)**

Twenty years after the tsunami, different types of reconstructed housing showed various long-term outcomes. Much of the "non-relocated" "provided" housing, built by the army on residents' land, was small and poorly ventilated, had been built quickly, and was in poor condition. On the other hand, professional builders had constructed the donor-driven "provided-relocated" housing. Although in the previous 10 years (the first decade after the tsunami) residents were less satisfied with their houses, 20 years after the tsunami residents' satisfaction with their houses had increased. In the case of self-built relocated houses, residents participated in design decisions; in self-building projects, residents worked with the volunteers to build their own houses. This involvement in the reconstruction process fostered residents' attachment to their housing and community over the long term.

### Cultural Responsiveness and Adaptation

In the self-built housing resettlement sites, the house design and spatial planning of the site, carried out with the participation of the community, gave residents the ability to expand their houses. The inclusion of second-floor spaces and spaces in front of the houses represent the cultural responsiveness of these houses and their potential for adaptation. In addition, two of the four community-based projects supported by CODI in Namkhem included culturally specific support—one for a community of fishermen, and one for an indigenous community.

For the Tok-Pu fishermen community, residents wanted to live on the water near a mangrove forest, and they were able to negotiate with the local government to receive half the standard reconstruction costs and self-build houses over the water on concrete pillars. For the Morgan indigenous community, supported by CODI in Phang Nga, their houses were designed and built based on incorporating cultural factors, including wooden building materials and elevated houses with open living spaces on the ground level. However, most of the houses they received still have some aspects that do not fit well with their culture of living nearby the sea, and they do not have enough space for fishery equipment in some areas.

### Redundancy and Integration

On the scale of settlement, new communities were created by the post-tsunami relocation projects, where housing was provided to residents who no longer wanted to live near the sea. The many residents still inhabiting houses in these resettlement sites decades later represent risk reduction for these communities. Preservation of mangroves also serves a redundancy function, as it mitigates future tsunami impact.

However, the national government took a longer time than expected for housing relocation, reconstruction, and the overall recovery process. Often, residents ended up choosing to live in hazard risk zones due to proximity to jobs, rather than choosing hazard safe locations.

### Rights, Equity, and Security

Local government worked together with volunteers and the military to support residents displaced by the tsunami and provided land and shelter for them until permanent house construction was finished. Local government was responsible for confirming the number of households and residents' ownership status, and surveyed residents about preferences regarding relocating to provided houses versus staying and rebuilding on their former land. With this process, housing reconstruction was implemented with equity for the survivors. In addition, through the support of CODI, some groups of residents (renters or fishermen living over the water), who were initially excluded from housing recovery support, were able to negotiate support.

## Case 2: 2011 Flood, Ayutthaya

The 2011 flood in central Thailand was a massive disaster, causing significant damage to 1.96 million households. Despite the government's significant spending on flood relief support, which totaled 45 billion Thai baht (THB) or 1.241 billion USD, the housing reconstruction outcome was insufficient and unsuccessful for supporting residents, as can be understood by considering the reconstruction situation and problems in flood relief support policy from residents' point of view (Fig. 3.4).

### *Disaster Impact*

The severity of the 2011 floods was due to heavy rain from five storms that hit north and central Thailand during the tropical monsoon season from June to August, compounded by geography. Water ran from the north down to the flat floodplain of central Thailand. Exceeding the capacity of rivers in 16 of Thailand's 76 provinces, water could not flow rapidly to the

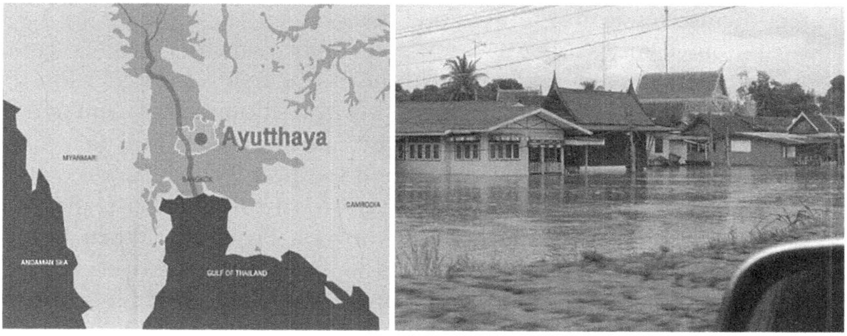


**Fig. 3.4** Flood-affected areas along the Chao Praya River, where flood waters remained for 1–3 months in Ayutthaya (left). Floods affected every house in Ayutthaya (right). (Source: T. Sararit)

sea, resulting in severe floods in 14 provinces in central Thailand, including Ayutthaya Province and some parts of Bangkok City, with flood waters that remained for several months. According to "2011 Thailand Floods: Event Recap Report-Impact Forecasting," 1.5 million homes and other structures were adversely impacted by the flooding (Aon Benfield, 2012; World Bank, 2012).

*Pre-disaster Housing Context of Ayutthaya, Thailand*

In Ayutthaya Province, floods are a common occurrence, lasting up to three months. Floods stop all economic activities; people lose income and/or jobs. Ayutthaya's economy relies on agriculture, industry, and tourism. As a UNESCO World Heritage site and famous tourist destination, most research on Ayutthaya's flood disasters focuses on flood-risk prevention and loss in industrial and economic sectors.

Studies of housing that is responsive to flood risk in the area have recommended two types of houses: "elevated" houses on stilts or pillars to mitigate flood impacts, or houses that float when the floodwater level exceeds the second floor. These studies suggest that housing renovation for flood mitigation is reliant on building structure, construction costs, and building integrity. However, beyond receiving financial support and learning from previous flood events, residents need encouragement and motivation to rebuild houses to higher standards to mitigate flood risk. It is crucial to understand the relationship between residents' economic factors and costs required for construction, as these factors can support residents' ability to choose the best treatment to renovate their houses and protect themselves against future flood disasters.

Most houses in Ayutthaya are single-family two-story houses built along rivers, or concrete single-family houses near roads and rice fields, in former agricultural communities currently transitioning into industrial areas. Typically, in central Thailand, most land in Ayutthaya is flat. The area along the Chao Praya River is a good location for rice paddies, and hence, agriculture is the main economy in surrounding provinces—Singburi, Ang Thong, Lopburi, and Ayutthaya Provinces. When heavy rain occurs, massive amounts of water flow from northern rivers to the Chao Praya River. In the central Thailand flood plain, water is released into a reservoir at the Pasaksholasit Dam; if it reaches capacity, water is released into lower regions of the Chao Praya River. This causes floods to happen regularly.

*Flood Damage and Government Policy*

Affecting 26 provinces, the 2011 flood was the most devastating disaster in Thailand in terms of affected area. It took three months for flood relief and caused a significant impact because of the timing and damage to the Thai economy, affecting not only households but also the agricultural and industrial sectors. Disaster damage to property primarily impacted housing, causing about 1 billion USD in damages. Even for less-impacted housing structures, a lot of furniture and work equipment were destroyed during three months of flood inundation. Cleaning took time and required financial assistance from the government and donors. Regardless of house type, flooding experience is common for residents of Ayutthaya, yet most residents continue to build their houses based on construction costs and space constraints without prioritizing disaster prevention. Even residents who experience floods every year or live in flood-risk areas still live in one-story houses because they cannot afford to build elevated or two-story houses. Even among residents whose families have lived for generations in Ayutthaya, a frequently inundated flood plain with many traditional Thai houses in the past, current residents are finding two-story, open, elevated houses or Thai traditional houses less attractive because of the high costs of construction and maintenance. This changing perception is due to changes in ways of living and lifestyle. Most houses in Ayutthaya Province are two-story concrete structures, with some half concrete (first floor) and half wood (second floor). In the 2011 floods, all houses were affected. After the water receded and flooded areas dried, housing damage levels were evaluated by the following categories.

1. The floor had to be cleaned, or floor tiles replaced because of mud and dirt.
2. Furniture inside houses was destroyed by water.
3. The watermark can be seen on every wall of the house.
4. The flood deposited a lot of debris in the community surrounding the house.
5. The water damaged the house's foundation.
6. Flood waters destroyed doors, frames, windows, and glass.

*Government Policies Related to Flood Relief*

Interviews with 80 residents from the six villages in Huntra sub-district revealed that the national government extended flood relief to all residents in the affected areas, providing emergency relief of 5000 THB (137 USD) per household.[1] After flood waters receded, the Sub-district Administrative Organization (SOA) sent staff members to estimate the damage incurred by individual households based on lists of their livelihood losses. Residents were then granted flood relief assistance up to 20,000 THB (550 USD), calculated using a percentage-based checklist of flood loss in eight categories: kitchenware (22%), building materials (27%), lighting (3%), clothes (13%), uniforms (5%), beds (7%), work equipment (13%), and other things (12%). However, despite the seemingly high percentage of flood relief assistance, most residents only received a small amount of money.

Subsidies residents received were insufficient, as evidenced by data from Village number 1 in Huntra. 479 residents (61.5%) out of 779 households in Village number 1 received only a very small amount of assistance, ranging from 13,000 to 14,999 THB (360 to 445 USD). For building materials, 43.5% of residents received a meager amount ranging from 1000 to 2999 THB (30–80 USD), which is not enough for renovation or reconstruction.

Ayutthaya residents' average income is 22,302 THB (615 USD) per family per month. During floods, buying food or beverages became difficult as consumer goods' prices rose 2–3 times. Equivalent to only one month's income, the national government's flood relief money (25,000 THB or 690 USD) per family was insufficient, as it covered only 22.84% of the actual amount residents lost, equated to 105,454 THB (2900 USD). Financial assistance for building materials was only 4078.5 THB (112 USD)–3.73% of total losses. Renovation expenses, based on surveys with residents, averaged 36,850 THB (1015 USD), and savings spent during the emergency period averaged about 24,000 THB (660 USD) over two months. In one case, owners used savings of 250,000 THB (6893 USD) to renovate their house, while support money was only about 5000 THB (138 USD).

Based on an average monthly income of 22,302 THB, over two months of flooding residents lost an average 44,604 THB in income. Moreover, as residents only received flood assistance six months after floodwaters had dried up (in May 2012), many needed to borrow money from banks or the private sector for house repairs before receiving flood relief money. After suffering from floods for three months, residents faced other problems, such as difficulty in renovating their houses back to habitable

[1] Exchange rate: 1000 THB = 30 USD

conditions or better protecting themselves from flood waters. As renters did not qualify for the above-mentioned assistance, only receiving the emergency relief of 5000 THB (138 USD), they faced additional challenges.

*Housing Repair, Renovation, Reconstruction, and Results*

With limited flood relief funding from the government and limited (material assistance) or no additional support for housing reconstruction, residents used various types of renovations for flood mitigation. A specific case study in the Huntra sub-district of Ayutthaya illustrates the ongoing challenges faced by communities in flood-prone areas. Huntra, located between the historical center of Ayutthaya and the Rojana industrial zone, consists of six villages connected by the Huntra canal. The area spans 7.22 square kilometers and is home to over 5000 people in more than 2000 households. The proximity of Huntra to the Chao Phraya River means that it experiences regular flooding, especially during periods when water is released from upstream dams. All areas of Huntra flooded during the 2011 flood and it took more than a month for the water to completely recede.

Based on interviews with 20 households in Huntra in 2012, renovation costs ranged from 10,000 to 100,000 THB (297 to 2970 USD), depending on building damage. Interviewed residents carried out the following four types of renovations. (1) Residents of two houses raised the second floor to create an open-air space on the ground floor level underneath; this was quite expensive. (2) One resident adapted their house by extending the second-floor area to have more rooms and a toilet on the second floor, saying that if floods occured again, they can live more conveniently on the second floor. (3) Owners of 12 houses chose to paint and/or prepare their houses rather than changing the building structure. These actions include repainting, buying new furniture, preparing a boat, and moving everything to an upper floor. (4) Residents of five houses did not mention any renovation or preparation, saying they did not have enough money, however they were still worried that floods could happen in 2012 or 2013.

**Relocation Intentions**

Considering they may be affected by large-scale flooding again, 40 Huntra residents were interviewed in 2023 about their intentions to relocate permanently away from Ayutthaya to another place. 25 residents (62%) were willing to stay in the same place even if they were at risk of future flood disasters because their work was in Ayutthaya. Ten residents (25%) planned on moving if the level of the next flood went higher than the second floor of their house, and five residents (13%) said they will move if they experienced a flood again.

*Long-Term Outcomes 12 Years After the Disaster*
With almost no structural renovations post-flood, discussion and evaluation of long-term housing outcomes focused on flood mitigation actions and behaviors. According to surveys, from 2011 to 2023 residents' flood awareness increased; in 2011 only 10%, or 4 out of 40 residents, responded that they were willing to prepare for flooding physically, such as locating electric outlets above water level or elevating or extending upper floors/ upper floor area. Compared to that, in 2023, 33 out of 40 residents said they would prepare for floods, but not by modifing housing structures; they prepare sandbags, elevate furniture, and place belongings on furniture to be above the water level. Most residents tended to move belongings to upper floors and prepared food and goods in advance of flood events, which could be more helpful in the long term. However, only a few residents said they would take action towards significant house renovations. Most residents agreed that information on flooding and water levels is essential. However, seven residents said that they did not care about flooding information, although they had already experienced flooding.

**Robustness and Diversity (Buffer Capacities)**
Of 40 households surveyed, only four had taken steps to rebuild their homes by elevating the floor or extending the second floor, which suggests there was limited robustness of rebuilt housing over the long term. It is crucial that residents enhance their homes' flood resistance to reduce risk exposure if they want to continue living in current locations. However, most residents are unable to do so due to the type of house structure and financial constraints.

**Cultural Responsiveness, Appropriateness, and Adaptability**
Although most residents were limited in their abilities to make substantial physical modifications to their houses, those who could raise their homes or extend second-floor areas demonstrated cultural responsiveness and adaptability on the part of the residents themselves, if not from official government support.

**Rights, Equity, and Security**
With relief assistance that covered only a small fraction of residents' losses and repair needs, and even less provided to renters, housing recovery lacked equity, and did not provide security for survivors. Inclusiveness is also needed from the flood disaster mitigation phase to enhance effectiveness of post-flood housing recovery, including the following two steps. First, for enhanced and localized data collection, systems should be implemented to gather detailed flood information at the local level, including water depth and other culturally relevant information such as religious preferences, income, rental, or ownership status. Secondly, vulnerable groups should be integrated into disaster management planning through the recovery phase.

*Summary of Housing Recovery Outcomes in Ayutthaya*
More than 12 years post-flood, residents had a greater awareness of flood-evacuation procedures, but only a few had managed to undertake significant physical structural measures for flood mitigation. The current focus on providing money to affected homeowners has been shown to be an ineffective flood relief policy. The government should provide a more structured approach, such as supplying construction materials, skills training, and assistance of craftsmen to help with renovations. More systematic support could enhance the community's resilience to future floods and improve the overall effectiveness of disaster recovery efforts.

## Comparative Analysis of Two Case Studies: Thailand

After giving an overview of similarities, differences, and changes in housing policy in Thailand after the two major disasters, this section explores the long-term recovery outcomes in terms of resilience characteristics, shown in Table 3.1. Finally, key factors are suggested that contribute to long-term resilient outcomes.

### *Similarities, Differences, and Changes in Housing Reconstruction Projects, over Time*

Both the 2004 tsunami and 2011 flood were national-level disasters because of the significant damage to affected households, the economy, and infrastructure, and the large number of deaths. Government management after both disasters was also similar, as the national government commanded each ministry or province in a top-down manner, and each ministry processed the funding for the local government for

**Table 3.1** Comparing resilience characteristics as long-term outcomes of two housing reconstruction projects in Thailand (●●● = strongly agree; ◎ = agree; ± = neutral; ⊠ disagree) (Source: T. Sararit)

| *Resilience characteristics* | *Resilience outcomes in housing systems over the long term* | *Description of long-term resilience outcomes in housing, communities, and institutions* | *Namkem Phang Nga 2004 tsunami (rural)* | *Ayutthaya 2011 floods (urban)* |
|---|---|---|---|---|
| Robustness and diversity (buffer capacities) | Planning and building standards | Well located for reduced exposure | ◎ | ⊠ |
| | | Building standards stronger/revised | ⊠ | ⊠ |
| | Structural safety | Structural hazard-safety features explicitly integrated in housing | ◎ | ◎ |
| | | House withstood another disruption | ⊠ | ⊠ |
| | | Good construction quality | ◎ | ◎ |
| | | Choice of materials and construction technologies | ◎ | ⊠ |
| | | Traditional/local construction labor incorporated | ●●● | ●●● |
| | | Replication of improved construction systems | ⊠ | ⊠ |
| Culturally appropriateness and adaptability | Housing design | Design with diverse housing typologies | ●●● | ± |
| | | Culturally responsive spatial layout | ●●● | ◎ |
| | | Thermally comfortable design | ± | ⊠ |
| | | Design flexibility for extension or adaptation | ●●● | ± |
| | | Occupancy in houses | ◎ | ◎ |
| Redundancy and integration | Settlement or city development | Services integrated with housing | ●●● | ●●● |
| | | Infrastructure integrated with settlement | ●●● | ●●● |
| | | Social amenities integrated in settlement | ●●● | ●●● |
| | | Livelihood proximity/considerations | ◎ | ◎ |
| | | Redundancy integrated in housing or settlement | ◎ | ◎ |

| | | | | |
|---|---|---|---|---|
| Rights, equity, and security | Equity, inclusion | Recovery for all residents | ◎ | ± |
| | Dignity, sense of ownership | Participatory process | ●●● | ◎ |
| | Affordability | Low cost, affordable, and ease of maintenance | ●●● | ◎ |
| | Security | Better security of land, housing tenure | ●●● | ●●● |
| Resourcefulness (adaptive capacities) | Institutional capacities: recovery program, assistance, agility, rapidity, timely, yet strategic | Enforcement of planning and building regulations | ◎ | ◎ |
| | | Improved governance and institutional capacities | ± | ± |
| | | Proactive investment in prevention strategies | ± | ± |
| | Social capacities | Changed social mindset, norm, and improved or preserved social bond | ●●● | ●●● |
| | Individual capacities | Improved household awareness, security, and reduced aid dependency | ●●● | ◎ |
| | | Better quality of life and political voice | ◎ | ◎ |

reconstruction. In comparing housing recovery policies in Thailand at the state level, not much has changed in 20 years. After the 2004 Indian Ocean Tsunami, and after the 2011 flood, most houses that were not completely destroyed were reconstructed by individual residents themselves, but owners of completely destroyed houses received support money from the government, and sometimes government support for reconstruction labor.

### *Role of National Government*

After both disasters, the national government took care of the reconstruction of basic infrastructure and rebuilding public buildings. A key difference is that after the 2004 tsunami, the national government was also directly in charge of housing reconstruction of permanent houses for survivors. Thailand's prime minister commanded direct control over six key ministries involved in recovery, which each oversaw the construction of permanent post-disaster houses on site. However, after the 2011 flood, the national government directly managed only the rebuilding of infrastructure and let the local government support housing recovery.

### *Role of Local Government*

After the tsunami, local government was responsible for confirming the number of affected households, residents' ownership status, and if they wanted to relocate to a provided house or stay on their former land. With less structural damage to houses in the 2011 floods compared to the tsunami, residents mostly sheltered in-house. Local governments supported relief operations (e.g. preparing food and distributing consumer goods) for communities in their area and were also responsible for qualifying residents for support money (150 USD) from the national government.

The two cases show the important role of local leaders and their impact on leading recovery policy. After the tsunami, residents negotiated and were involved in housing recovery decisions, especially important for most low-income residents, who could get support from NGOs and CODI for their permanent houses. However, limited support after the flood meant that residents could not carry out reconstruction.

*Robustness and Diversity (Buffer Capacities)*

After the disasters, residents moved from temporary to permanent houses in various locations due to budget and job factors, and sometimes with considerations of future disaster risk. Household damage level was a significant decision factor for relocation after the 2004 tsunami, which completely damaged more than 68.7% of houses in the tsunami-affected area. The shock impact, high rate of complete housing damage, and risk of future disasters led many residents to decide to relocate by choosing to move far away from the sea.

In contrast, after the 2011 floods, only 1% of houses were completely damaged. For two to three months after the 2011 flood, residents lived in shelters, on the second floor of their houses, or in self-made temporary shelters on the local road close to their homes.

Relocating was difficult in the case of the flood since most residents worked in agriculture, meaning their income was generated from their land that was located within the disaster-affected area. Conversely, most tsunami-affected residents worked in the fishing industry, and when relocating to a new place, some residents changed their jobs to work as drivers and laborers in the tourist business.

As the building quality of relocated houses is better than that of provided non-relocated houses that were rapidly constructed by the army, residents wanted to and continue to inhabit these relocated houses.

Since there was limited housing support and reconstruction after the 2011 flood, with most residents opting to repaint their houses rather than undertake behavioral flood mitigation actions, there is minimal evidence for housing robustness over the long term from this case. Looking at these two cases also shows how post-tsunami reconstruction may have been a one-time unique response to a mega-disaster, but did not translate to supporting subsequent post-flood reconstruction programs.

*Cultural Appropriateness and Adaptability*

One significant change in long-term social and cultural outcomes is that the community made strong bonds to prepare for disasters and create good practices for other communities. This included respecting indigenous knowledge in the reconstruction, which was observed 10 years after the tsunami; however 20 years after the tsunami, the cultural center of the Morgan ethnic group suffered from a lack of maintenance. New generations of indigenous residents now blend in with other local Namkhem people; however housing that was provided to them still supports them for the fishery business and is suitable for their way of life, evident in the members of this community continuing to inhabit the houses they received from the local government.

In both cases in Thailand, relationships between community members became stronger after experiencing the disaster together; this also became a key to good recovery. People in local communities support each other, also helping to participate in future disaster prevention activities. They help each other during floods and help each other clean or repair houses; sometimes, they are also able to share some of their experiences, and also benefit from the knowledge of laborers who can apply lessons from previously acquired experience with flooding to their craftsmanship and house repair work.

*Redundancy and Integration*

In both cases, post 2011 floods and post 2004 tsunami, long-term redundancy and integration outcomes are mixed. While hazard risk maps existed and were used in the immediate aftermath of the devastation, particularly after the 2004 tsunami, they are not used anymore to inform land use planning, and many new developments such as residential houses and hotels are now located in the tsunami risk area. With limited reconstruction and implementation of mitigation measures, there are also limited redundancy factors visible after the flood disaster.

*Rights, Equity, and Security*

Long-term equity outcomes vary across communities affected by different hazard events. Residents affected by the 2004 tsunami received differential assistance based on damage assessment, from the government as well as donors and NGOs. Moreover, even low-income communities received land and housing assistance from CODI, whereby they could achieve land ownership after 15 years. However, communities affected by the 2011 floods received uniform financial assistance insufficient for rebuilding, and hence most residents ended up merely refurbishing their houses (e.g. repainting or buying furniture).

Thailand's case highlights the differential treatment for reconstruction support following floods in comparison to tsunamis, or for slow-onset hazards compared to rapid-onset disasters. Comparing financial support for houses provided to residents by the government, the amount was almost the same after the 2004 tsunami and 2011 flood. Comparing only the percentage of houses destroyed, it may seem like the 2004 tsunami caused more damage. However, in terms of economic losses and security for residents, the 2011 flood also severely impacted residents' lives, and survivors also needed new or safer houses to mitigate damage for future disasters. The national government provided funding for victims of the 2011 flood as "comfort money" of only about 5000 THB per family (150 USD). As this was not enough to rebuild their houses, they could only "paint back better."

*Resourcefulness (Adaptive Capacities)*
In the long term, the capacity of communities to make informed decisions for their housing extensions were supported by CSOs in both cases. CODI has now accumulated more experience, over 20 years, in supporting housing for low-income communities, not only after disasters but also in support of low-income community development in non-disaster situations, with funding of 242 million baht (7.1 million USD) per year.

Long-term capacities of the national government increased from the 2004 tsunami to the 2011 floods, and awareness of the government was applied to planning and preparing for disaster risk by providing emergency funding for disaster relief. However, if a disaster occurs at the local level, it will be managed by the local authority.

## Key Factors with the Potential to Catalyze Resilient Long-Term Outcomes

From a comparison of two Thai case studies, we identified the following key factors that have the potential to lead to resilient long-term outcomes.

### *Participatory Process Involving Multiple Sectors, Communities, and Authorities*

*Role of CODI*
The Community Organizations Development Institute, or CODI, is vital in Thailand's housing recovery process. Working with community networks across Thailand to turn disaster events into development opportunities, CODI helps residents plan, manage, and adapt their relief and rehabilitation (CODI, 2005). Specifically regarding flexible management for post-disaster housing reconstruction and related loans, the 2004 tsunami was CODI's first project to focus on supporting post-disaster housing recovery; previously, CODI's housing projects had primarily supported low-income residents (ACHR, 2005). CODI provided various types of reconstruction support, including self-built relocated houses, and

continued to support residents over the long term. Using a community mortgage, CODI provided financial support for residents' groups to buy land, which they own after 15 years. For non-relocated house projects, CODI participated in planning and brainstorming activities and allowed residents to participate in reconstruction and decision-making processes. CODI was not the main supporting organization after the 2011 flood, when local governments and NPOs were the main organizations supporting residents.

*The Roles of NGOs and Community Networks*
Non-government organizations' contribution to disaster reconstruction in Thailand is also significant, as these organizations have more funding flexibility and can work faster and more closely with community networks and/or residents than the government. After the 2004 tsunami, many NGOs came to southern Thailand to support the recovery, not only by providing funding but sometimes by designing post-disaster houses, and/or providing support for building materials as well as planning. The quality of housing provided by NGOs was better than those constructed by the military. NGOs not only supported building houses but also continued to support other aspects of residents' recovery process, such as funding for extending houses or providing support for education and/or safety training for the local community. After the 2011 flood, the Red Cross and other NGOs also supported residents with food and water, and through their projects, many donations from across Thailand and other countries supported infrastructure reconstruction.

Along with non-government and/or non-profit organizations (NPOs), community networks also played an important role in recovery. Based on neighborhood groups or job-related networks, informal institutional arrangements can generate creative strategies for post-disaster recovery (Kamel, 2010). The community network of Namkhem is one example, where residents gathered initially because they were promised that boats would be provided for fishermen who lost their boats in the tsunami. From there, residents formed a group and with the help of CODI, they negotiated for their houses not to be relocated to another area, since their fish farming and crab fishing required them to stay near the shore. These networks in crisis worked with CODI to find solutions and channel funds to a reconstruction process. After the 2011 flood as well, many networks operated to support residents through activities such as sharing boats, food, and consumer goods. In 2023, social media also played an important role in flood warning and information sharing between residents such as the Ayutthaya city page on Facebook or LINE groups that share information about available help within residents' communities.

### *Residents' Participation and Ownership of Post-disaster Housing Reconstruction*

After the 2011 flood, there was less resident participation in housing reconstruction, decisions, or future flood plans; most official recovery plans and discussions focused on dealing with economic risk for industrial areas and entrepreneurs. For this purpose, the national government also supported future flood protection systems, but there were no public hearings for housing recovery. Public hearings, during which policies are announced to citizens with transparency, are part of standard procedure for introducing new policies such as the designation of hazard area or housing recovery programs. Without public hearings for post-flood policy, the national government indicated floodwater collection areas, designated for flooding to protect other areas in the industrial zone, without discussion with local residents. Without any plan for evacuation or support for residents who survived and may have their rice fields in that area, residents cannot adapt their houses to be higher than water levels or prepare better to withstand future floods. Instead, most of the official money for floor recovery went to infrastructure such as dams, water gates, and flood walls.

After the 2004 tsunami, residents in the case study area mostly worked in the fishery industry, and as many of them lived in a slum near the shore without house ownership pre-tsunami, they could not access the standard support provided to pre-tsunami homeowners. However, as CODI supported them, together with international NGOs, these residents have participated in many processes. Starting from gathering as a community network, they worked out which locations were safe and had available land. Through the community mortgage project that was set up to rebuild their housing, every month residents make payments to CODI, and after 15 years (i.e. in 2019) they received the land rights and ownership. Until then, their house ownership rights allowed them to sell the house to others and adapt their own house.

People who had land ownership had more opportunities for post-disaster reconstruction support, including (1) receiving a provided house, built by the military; (2) getting a house provided by a donor in a new relocation area; or (3) receiving support money from the government and rebuilding a new house on their own land. Each of these options included a different level of resident participation. Interviews with residents confirmed that residents who were more involved in the recovery processes also felt more strongly connected to their community and were more satisfied with their own houses.

In both Thailand's case studies, the local government played an important role in the implementation of housing reconstruction, not only by building houses, but by local leaders being key people to connect residents and create activities that lead to better recovery. The key factor was how they could gather opinions from local residents and share them with the government to find common ground between people's needs and the implementation of limited available support from the government.

After 20 years, some members of local government are still active in efforts to prevent damage from future disasters, and also continue to be important persons to connect people in the community together through activities. In addition, good relationships between neighborhood residents have created good environments, help keep the community alive, and support survivors in new chapters of life such as new jobs. Residents have been able to extend their houses to express their own character and also rebuild or repair their houses over time.

### *Neighborhood or Settlement-Based Planning*

The neighborhood and settlement planning factors reflect their importance as years pass. After the 2011 floods, the settlement remained the same. However, residents who experienced the disaster saw the importance of connection and bonds in the neighborhood, assisting each other during floods and contacting each other for sharing flood information and warnings.

After the tsunami of 2004, at the beginning of the housing reconstruction process, people who had gotten a house first lived on their former land, the same as in pre-disaster times. They were initially more satisfied than the relocated group. For the groups that relocated, the support after the 2004 tsunami was unique, provided to a variety of residents and also from various donors. However, the groups of residents who moved to new places together created better quality houses because they used professional builders, and while they lived on government land, the settlement included infrastructure. Two years later, they were also provided with additional funding to extend and renovate the relocated house according to the their needs.

At the same time, relocation projects incorporating self-built houses also had good levels of residents' participation in reconstruction, and created strong neighborhood connections. They have shared land, and built houses together with help and support from CODI; the quality of houses is good, and the community worked together for paying their community mortgage every month for 15 years (until 2019). The ownership of land and house has been transferred to the residents, and they continue to live in the same houses, which have not been sold or left vacant. Economic activities of Namkhem community have changed from fishery to tourism. The location of houses far from the sea is no longer a problem, because residents have adapted their houses, livelihoods and lifestyle to suit their new lifestyle. For example, they have renovated the houses to add garages to park cars used for tourist business or have built more rooms for extended family members.

### *Flexible Design Offering Adaptability over Time*

From a housing recovery perspective, the case of Namkhem demonstrates the positive aspects of flexible housing reconstruction choices. While reconstruction projects initially resulted in less satisfaction with relocated houses, ongoing support and funding for individual houses' spatial extensions made a significant difference. The ability to extend their living spaces helped relocated residents adapt to their new jobs and lifestyles, demonstrating their resilience and ability to move on after the disaster. Conversely, housing that does not allow for self-extension may lead to less satisfaction among residents.

Having more options from the emergency shelter phase also supports residents' flexibility. With more flexible sheltering options, residents have more choices and more room, and they can feel secure when they are able to have shelter near their house or can take shelter on the upper floor of their house. Although there was minimal physical housing reconstruction for flood-affected residents, most adaptations to housing in the flood-affected area are very flexible, elevating floors according to the water levels from previous flood experiences. Residents made adaptations in consideration of how to make their house withstand the next flood, or at least how they will be able to stay in the upper floors of the house during floods that often last more than a month.

## Conclusion

Located in a region prone to natural hazards such as floods, earthquakes, and tsunamis, Thailand has had its fair share of disasters, with the most notable being the 2004 tsunami and the 2011 flood. These disasters had a significant impact on the country, particularly on the housing sector. As a result, the Thai government has developed a housing recovery policy that relies on local authorities and regional leaders to a great extent. The flexibility of the housing recovery process in Thailand is unique, with negotiations between residents, local governments, and non-profit organizations often playing a significant role in the process.

In large-scale disasters or those with extensive damage, a variety of housing projects may be required. While the government rebuilds infrastructure, self- or owner-driven housing is a crucial part of recovery. In Thailand, many housing recovery projects involved residents themselves in the design and building of their homes. Local laborers are typically involved in construction, but there is a growing need for greater awareness of climate, green building practices, and cultural considerations.

The development and implementation of zoning and early warning systems has been a priority since the two major disasters in 2011 and 2004. However, there is still room for improvement in the quality of buildings and individual house types. Despite future risks, residents tend to stay put, making it even more important for communities to work together to prevent future disasters and participate in the housing recovery process.

In conclusion, the case studies from Thailand provide insights into the impact of major disasters on an upper-middle-income country and the long-term recovery efforts required for affected communities. The 2004 tsunami and the 2011 flood highlighted the need for better disaster management and prevention policies in Thailand. Although the government provided partial financial support for flood relief, the outcome of post-flood housing development was not as successful as it could have been. Understanding the reconstruction situation and problems in flood relief support policy from the residents' point of view is crucial for successful long-term recovery. The study of the 2011 flood in Ayutthaya Province demonstrated that floods are a common occurrence in central Thailand, and they can last for months, causing significant damage to millions of households.

On the other hand, the case study of the 2004 tsunami in Thailand provides examples of diverse housing recovery approaches that include many stakeholders, and range from central government driven to community-driven housing recovery projects. This helps fulfill residents' basic needs and improve the quality of their homes. Additionally, participation of the community is important for future disaster management and prevention policies. By working together with residents, local and central government, NPOs, and other stakeholders, communities can be better equipped to cope with future disasters and recover more quickly and effectively in their aftermath. The type of disaster and affected area (rural or urban), and extent of housing damage can make a difference in the speed of recovery. Regardless, they share the same imperative for housing recovery policy. The experience from Thailand serves as one example of how community-driven initiatives can play a crucial role in disaster recovery efforts.

## References


Asian Coalition for Housing Rights (ACHR). (2005, August). *Tsunami Newsletter* (pp. 32–40). Community Organizations Development Institute (CODI). Retrieved from https://en.codi.or.th/wp-content/uploads/2019/11/Tsunami-Newsletter-ACHR-Aug-2005.pdf

Benfield, A. (2012). *2011 Thailand Floods: Event Recap Report-Impact Forecasting*. March.

Kamel, N. (2010). Lessons for Long-Term Residential Recovery: Factors of Community Resilience and Marginalization. *Journal of Disaster Research, 5*(2), 147–154.

Maitree, J. (2013). *Namkhem Community DDPM Volunteer Presentation* (pp. 1–40).

Sararit, T., & Kondo, T. (2014). Housing Renovation after the 2011 Thailand Flood in Ayutthaya. *Journal of Disaster Research, 9*(4), 563–570. https://doi.org/10.20965/jdr.2014.p0563

World Bank. (2012). Thai Flood 2011 Rapid Assessment for Resilient Recovery and Reconstruction Planning.

CHAPTER 4

# Long Term Post-Disaster Housing Reconstruction in Japan: Case Studies (High-Income Country)

**Abstract** Exposed to various natural hazard risks, Japan has a long history of disaster experience, established governance structures, and precedents for disaster management, response, and recovery. In the last 75 years since World War II, Japan became a highly modernized and industrialized country, incorporating strong disaster management along with economic development, and even promoting its experience with economic growth and disaster risk reduction as part of foreign assistance and disaster diplomacy in Asia (Maly, 2025) Although Japan's economy is no longer growing, it retains strong governance and financial commitments to disaster mitigation, response, and recovery as a high-income country. However, disasters in Japan cause severe human and economic damage, including after the two cases in this chapter: the 1995 Great Hanshin Awaji (Kobe) Earthquake and 2011 Great East Japan Earthquake, tsunami, and nuclear disaster. Looking back at these two examples, 30 and 14 years ago, respectively, this chapter considers the long-term outcomes of post-disaster housing reconstruction. Drawing on past research, information was compiled from multiple sources, including reference materials and site visits. To clarify experiences and opinions about long-term outcomes, semi-structured interviews were conducted with recovery stakeholders, including current and former government staff, academic experts, community planners, non-profit organizations, community leaders, and residents. Between January and May 2024, interviews were conducted with 10 stakeholders in Shichigahama Town and 10 stakeholders in Kobe



M. Vahanvati et al., *Resilient Recovery from Disasters*,
https://doi.org/10.1007/978-981-97-8047-1_4

(Fig. 4.1). Following an explanation of Japanese contexts of hazards, disaster management, and reconstruction policy, analysis of each case considers the disaster impacts, housing recovery policy, and implementation. Looking back at the outcomes, we can understand the continuity and commonalities of post-disaster housing recovery in Japan, and the evolution of housing recovery since 1995. The conclusion reflects on strengths and weaknesses of Japanese government-driven housing recovery, suggestions for improvements, and potential lessons for other countries related to roles of government, CSOs, and non-profit organizations, holistic recovery, and community empowerment for risk reduction and resilience.

**Keywords** Japan • People-centered • Post disaster • Housing reconstruction • Disaster resilience • Long-term recovery • Great East Japan Earthquake • Kobe Earthquake

## Hazard and Disaster Context

Located on the Pacific Rim of Fire, where four seismic plates meet, the island archipelago of Japan is exposed to various natural hazards. With a long history of earthquakes, tsunamis, and volcanoes, 18.5% (302) of all earthquakes of magnitude 6.0 or greater between 2004 and 2013 occurred in Japan (Cabinet Office, 2014). Japan also experiences regular typhoons,

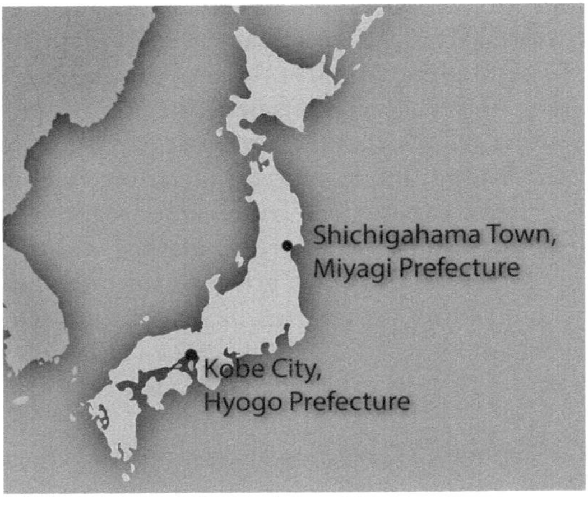


**Fig. 4.1** Map of Japan Case Studies. (Source: Authors)

heavy rains, floods, landslides, and heavy snowfall, and climate change has caused more severe meteorological hazard impacts in recent years.

Historically, dense low-rise settlements of wooden buildings made fire a serious and reoccurring risk in Japanese towns. Fire caused the majority of damage in Tokyo after the 1923 Great Kanto Earthquake, and 20 years later after devastating firebombing in World War II. Although war-time disasters received less attention, Japan suffered typhoon, earthquake, and tsunami disasters before and during World War II. During post-war reconstruction and early stages of economic growth, development, and technological progress, Japan's modern disaster countermeasures took shape in the late 1950s.

Early laws included the 1947 Disaster Relief Act after the 1946 Nankai Earthquake; the 1949 Flood Control Act after 1945 Typhoon Ida and 1947 Typhoon Kathleen; and the 1950 Building Standards Act (building code) established after the 1948 Fukui Earthquake. A turning point, immense damage during the 1959 Ise Bay Typhoon led to the establishment of Japan's 1961 Disaster Countermeasures Basic Act (Cabinet Office, 2014), which created the national government's Central Disaster Management Council and the Basic Disaster Management Plan, with a focus on clear national government responsibilities and disaster prevention structures (Cabinet Office, 2014).

## Disaster Management and Housing Context

Japan's national government plays a centralized and dominant role in all aspects of disaster management and recovery policy. Continuing the logic of urban redevelopment after World War II, the focus of government policies (and funding) supporting the reconstruction of disaster-affected areas remains on infrastructure improvement for economic growth, although Japan has been facing a shrinking economy and declining population for decades.

Reconstruction policies after the 1995 Great Hanshin Awaji (Kobe) Earthquake, the first major urban disaster in Japan since World War II, set the direction for Japan's subsequent recovery approaches over the last three decades. As Hirayama (2000) explained, Japan's housing policy is a "two-tiered" system: people are encouraged to acquire their own houses in the private market and public housing is provided for those excluded from the private housing market. With a similar approach, government support after the 1995 earthquake focused on the provision of public housing for the most economically vulnerable, rather than direct housing reconstruction support for homeowners. However, this resulted in social and spatial isolation and negative impacts, especially for elderly earthquake survivors (Hirayama, 2000). Revised several years after the Kobe Earthquake, legal policy in the 1998 Act on Support for Reconstructing Livelihoods of Disaster Victims now allows for more support for private homeowners, up to 3,000,000 yen (approx. 21,000 USD at that time) for households with completely damaged houses (Cabinet Office, 2014). However, the underlying logic for Japanese official support remains: government funds (raised from taxpayers) should not be used to replace private property. Although recovery projects after the 2011 Great East Japan Earthquake (GEJE) support individuals' housing reconstruction, this is conceptualized as supporting the community as a whole, rather than individual homeowners. For example, Collective Relocation for Disaster Mitigation, a project used extensively for residential relocation and recovery in tsunami-affected municipalities, provides new residential lots, including infrastructure and site development.

With a long history in Japan, wooden housing construction is still common for single-family homes. Multi-family and multi-story housing is also common in urban and suburban areas. Japan has one of the oldest building codes with seismic regulations, established in 1924 after the 1923 Great Kanto Earthquake. Since 1981, modern seismic codes have been part of Japan's Building Standard Law. Older buildings may not meet current standards, but construction quality is very high, and buildings constructed after 1981 comply with these codes. Therefore, few disaster deaths and injuries are caused by structures that cannot withstand earthquakes or poor construction quality of new buildings. However, people can be killed or injured by falling furniture, collapse of older buildings or block walls, fires, or tsunamis.

## Case 1: Housing Reconstruction after the 1995 Great Hanshin Awaji Earthquake, Kobe City

Japan's first major urban disaster post World War II, the Great Hanshin Awaji Earthquake struck Kobe City and Hyogo Prefecture at 5:46 am on January 17, 1995. Kobe's housing reconstruction policies shaped subsequent Japanese housing recovery programs. The Kobe Earthquake became known as the year of the "birth of volunteerism" in Japan, with many volunteers and non-profit organizations supporting disaster survivors and affected areas. However, housing recovery in Japan was (and continues to be) almost exclusively government-driven. With financial support from the national government, after emergency evacuation, prefabricated temporary housing units were provided, followed by primarily disaster recovery public housing, in the form of apartments in multi-story buildings, for low-income disaster survivors as permanent housing recovery support.

### *Pre-disaster Housing Context of Kobe, Japan*

Thirty minutes by train from Osaka, western Japan's metropolitan economic center, and an hour from the historic capital Kyoto, Kobe City is Hyogo Prefecture's largest municipality. A port town with centuries of international trade, Kobe is famous for westernized buildings, its port, and Chinatown. Geographically, Kobe City is long and narrow running east to west, and locals navigate by the Rokko mountain range (north) and the sea (south). Eastern Kobe, an easy commute to Osaka, was/is home to more well-off residents, with more working-class communities concentrated in Kobe's western areas.

Large areas of Kobe that were destroyed by U.S. firebombing in World War II were redeveloped and modernized in post-war urban reconstruction projects. However, areas that escaped World War II bombing—densely built, low-rise neighborhoods, with traditional wooden rowhouses and narrow alleyways—suffered the heaviest earthquake damage on January 17, 1995. These areas in western Kobe, with old and therefore more affordable housing, including small-scale factories, were destroyed primarily by fires that burned for several days, as rubble blocked narrow streets and emergency vehicle access.

The earthquake fault that ruptured passed under Kobe City and Awaji Island, also in Hyogo Prefecture. South of Kobe, in the Seto Inland Sea, Awaji Island is known for its peaceful farmland. Awaji and other smaller municipalities in Hyogo Prefecture also suffered housing damage, and today a portion of the fault and an earthquake-damaged house is preserved in the Awaji Earthquake Museum.

### *Earthquake Damage and Government Policy*

The magnitude 7.2 Great Hanshin Awaji Earthquake killed over 6400 people and destroyed over 240,000 houses. Neighborhoods of older wooden buildings in western Kobe suffered heavy damage due to fires. With some of the most affordable housing stock, these areas housed many lower-income and working-class residents, including immigrant communities from China, Korea, and Vietnam, and workers of small-scale factories, such as those fabricating rubber shoes. Most victims were killed by falling objects or furniture, crushed under rubble, or trapped and unable to escape fires. With narrow alleys blocked by debris, rescue workers and firefighters couldn't enter devastated areas; most survivors were rescued by neighbors.

*Large-Scale and Small-Scale Recovery Policies*
Kobe's recovery was a two-step process; the first step was the recovery plan created by Kobe City and the second step was the revision/modification of these plans in a process that included the involvement of local community development or *machizukuri* organizations (Be Kobe, 2015). Recovery policies and programs included land readjustment and large-scale redevelopment projects, community-based participatory planning, and housing support provided by the government primarily in the form of disaster recovery public housing.

*Targeted Redevelopment by Zones*
Based on damage levels, Kobe City was categorized into "black" (most heavily damaged), "gray" (somewhat damaged), or "white" (least damaged/undamaged) areas. Reconstruction projects targeted "black" areas, including several large-scale redevelopment projects with residential and commercial buildings, parks, and amenities (Fig. 4.2). Reconstruction projects included street widening and land readjustment, in which owners give up some land for infrastructure improvements, or land is swapped within the project. Historically, land readjustment projects were part of Japan's urban upgrading and modernization of roads and infrastructure especially after World War II. Kobe's land readjustment projects created wider streets, rationalized land parcels, and revised residential lots for former landowners. In addition to reduced residential lot sizes, building restrictions limiting lot coverage for new construction created additional challenges for rebuilding, even if residents could afford construction costs.

*Recovery* "Machizukuri" *Community Development in Town Planning*
The Japanese word *machizukuri,* literally "town-making" (*machi* means town and *tsukuri* means making) refers to various activities including community-building, community development, or town or community planning. Emerging in the 1960s, the *machizukuri* movement's goal was improving the built environment, especially against pollution. Satoh

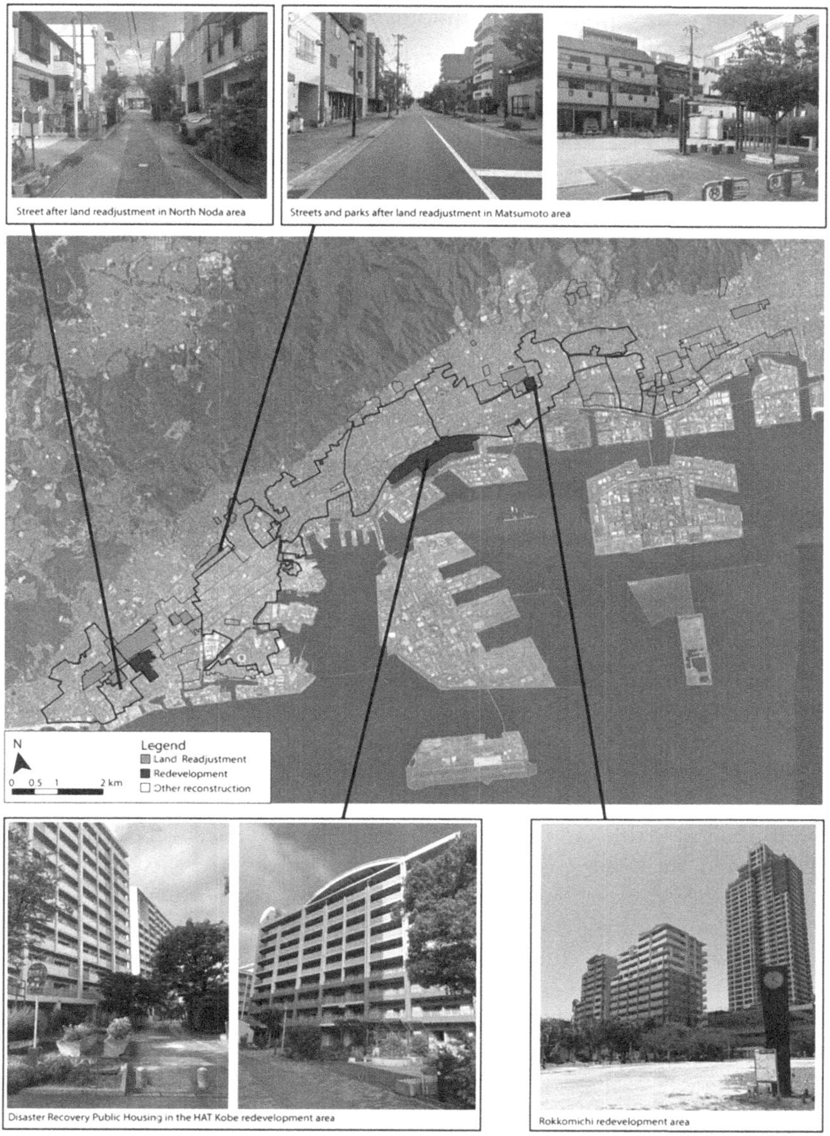


**Fig. 4.2** Housing recovery and targeted reconstruction areas in Kobe after the 1995 Great Hanshin Awaji Earthquake (Source: E. Maly, based on data from Kobe City)

(2019) and others pointed out unique aspects of Japanese *machizukuri* and meanings for community planning and community-building. Along with the Setagaya area of Tokyo, Kobe was one of the first cities to adopt a "Machizukuri Ordinance" in 1980, following local activism for community improvement and against pollution. The ordinance required that urban planning projects in Kobe City included a *machizukuri* process of citizen involvement; many Kobe planners and some communities therefore already had *machizukuri* experience before the 1995 earthquake. Drawing from the expertise of local *machizukuri* planners, Machizukuri Councils were established for each area of Kobe with earthquake reconstruction projects (Satoh, 2019) and played an active role in recovery.

### *Housing Reconstruction Plans and Implementation*

*Japan's Three Phases of Housing Support after the Kobe Earthquake*
The government provided support for three housing relief and recovery phases, established by Japan's legal framework for disaster survivor assistance. In the first phase, emergency shelters were set up, often in schools or other municipal buildings. Managed by local municipal governments, 220,000 survivors stayed in 599 emergency shelters (Hirayama, 2000). In the second phase, the prefectural government contracted the provision of temporary housing; in Kobe, these were prefabricated housing units. The national government provided funds to build and manage 48,500 temporary housing units; 30,000 were completed 3 months after the earthquake in March 1995, with the remaining units completed by August 1995 (Edgington, 2010). With the large amount of temporary housing needed and limited available land, mass temporary housing sites were created on land owned by Kobe City, many on sites far from the city center and evacuees' former neighborhoods. This weakened residents' ties to neighbors and community, especially as temporary housing units were allocated by lottery, prioritizing elderly or vulnerable residents, scattering survivors. For the third phase, government support for permanent housing recovery occured primarily through the provision of disaster recovery public housing for disaster survivors who were unable to rebuild their own houses.

*Japanese Government Support for Permanent Housing Recovery*
Although laws have been modified subsequently, the underlying logic of Japanese housing recovery support is that government money (gathered

from taxpayers) should not be used to replace residents' loss of personal private property. Instead, official support focuses on providing housing for survivors with greater need; this approach prioritizes social welfare over private property. Housing recovery policies in Kobe directly reflected this, focusing on the construction of disaster recovery public housing and providing very limited support for private housing reconstruction. In 1998, the updated Act on Support for Reconstructing Livelihoods of Disaster Victims (*Hisaisha Seikatsu Shien Ho*) provides up to 3,000,000 yen (20,780 USD in 2024) for households with totally damaged houses, which can be used for housing reconstruction (Maki, 2007). This policy is still in place today, with regional governments in some cases providing additional support.

*Two-Track Housing Recovery*

Japan's two-track approach to housing–primarily relying on the private housing market and avoiding private housing reconstruction support, with limited government support for those excluded from the private housing market, was reflected in Kobe's housing reconstruction. In the first track, those who could, rebuilt their houses using the private housing market or bank loans/savings, on their former or a similar lot, after land readjustment projects. There were also collective housing reconstruction projects, in which groups of residents joined together to rebuild new multi-family residential buildings, using subsidies for these projects made available for earthquake recovery.

In the second track, earthquake survivors unable to rebuild on their own could move into newly constructed disaster recovery public housing which was the primary focus of official housing recovery support provided by the government. Disaster recovery public housing in Kobe took the form of apartments in large multi-story reinforced concrete buildings; like all public housing in Japan, these were government-subsidized income-sensitive rental housing. Like temporary housing, many public housing sites were far from former communities; other public housing buildings were included in larger redevelopment projects. With disaster recovery public housing also allocated by lottery, connections with neighbors were severed again. Social isolation remained an ongoing problem, including solitary death, where residents pass away without any social connections, and their death is not noticed until later.

*Kobe's Housing Reconstruction Plan, Implementation, and Results*
In July 1995, Kobe City released the Emergency Three-Year Plan for Housing Reconstruction, with 72,000 new housing units planned (Hirayama, 2000), including 10,000 units of public housing for low-income residents, 30,500 publicly subsidized houses (semi-public, replacement housing for redevelopment projects), and moderate-income private rental housing (Hirayama, 2000). In July 1996, the Kobe Housing Restoration Plan increased the number of public housing units from 10,000 to 16,000, based on the temporary housing situation and needs of low-income and elderly survivors, while the total number of housing units in the plan remained at 72,000 (Hirayama, 2000).

In terms of units lost vs. rebuilt, the overall number of housing units was recovered. However, although 24,512 new public housing units were built in Kobe, the number built in the most severely damaged areas was far from replacing the 30,642 units of affordable low-rise wooden housing that had been lost (Olshansky et al., 2005). In Nagata Ward, one of the most heavily damaged areas, 11,711 houses had been lost but only 3171 units of public housing were provided (Olshansky et al., 2005). Not all survivors were able to return to Kobe after the earthquake; new housing was not necessarily provided in the same areas that had lost large numbers of housing, and even heavily subsidized public housing required some money for moving in, which made it out of reach for some residents (Murosaki, 2007).

### *Long-Term Outcomes 30 Years Since the Disaster*

Economic functions of Kobe's port disrupted by the earthquake never recovered, as alternative international shipping routes became permanent. In addition, as Japan's bubble economy deflated by the late 1990s, recovery did not occur within the optimistic pre-earthquake economic zeitgeist. Eastern parts of Kobe closer to Osaka were able to recover more quickly, as new private and public housing units were in demand in redevelopment projects in areas of Kobe such as Rokkomichi (Fig. 4.2, lower right photo). Yet in western Kobe, redevelopment projects such as in the Shinnagata area were planned with expectations of economic growth that did not occur, leaving vacant shops in multi-story shopping arcades.

*Housing Recovery Types (Paths) for Earthquake Survivors*
Long-term housing recovery outcomes in Kobe City varied by neighborhood and residents' reconstruction paths and project types. Post-earthquake housing recovery paths included (1) residents who rebuilt their own houses on former lots, in areas without government reconstruction or land use readjustment projects; (2) residents who rebuilt their own houses on new or modified lots created after land readjustment projects; (3) residents of cooperative housing redevelopment projects, where homeowners worked together to rebuild new multi-family housing on their pooled land; and (4) residents who moved into disaster recovery public housing. While long-term outcomes varied, overall, key aspects and comparisons to pre-earthquake conditions can be summarized as follows.

*Robustness and Diversity*
Regardless of housing reconstruction path or type, all new housing was built to current building standards. With high regulation compliance, these new structures should be able to withstand subsequent earthquakes or hazards. Government-provided public housing units and reinforced concrete multi-family structures were stronger than pre-earthquake housing, especially for former residents of older and lower-quality houses that were more likely to have suffered heavy earthquake damage.

*Cultural Appropriateness, and Adaptability*
Similarly, newly rebuilt housing represented a modernization of housing and living environments. Traditional streetscapes with narrow alleyways and dense communities of low-rise wooden row housing in the heaviest earthquake-damaged areas were replaced with wider streets, and in some cases, multi-story housing developments. While new and modern housing may be more convenient, these redevelopment patterns affected traditional ways of living. Many residents of single-story rowhouses, who interacted with neighbors daily pre-earthquake, were displaced from former communities and isolated after moving into disaster recovery public housing. In these new high-rise apartment buildings, provided as fully built, multi-family apartment-style housing units, public housing was not adaptable. Among residents who rebuilt their own houses, there was also little adaptation of modern housing units after inhabitation.

*Redundancy and Integration*

At the center of post-disaster reconstruction and redevelopment programs, urban planning projects included land readjustment (standardization and reorganization of properties and street widening) and related investments in roads and infrastructure. Kobe's recovery planning and redevelopment projects created many neighborhood parks and open spaces. While consideration of redundancies such as nature-based solutions or green reconstruction was very limited, this may not be surprising as 30 years ago these ideas were far from the mainstream.

*Rights, Equity, and Security*

The focus of official government support on providing disaster recovery public housing represents an idea of equity, prioritizing help for those who need it most, and affordability through subsidized rents. However, this focus also resulted in a large group of survivors, who might have been able to rebuild their houses with just a small amount of assistance, unable to rebuild and left with the sole option of moving into public housing. In addition, as housing was not replaced in the same areas and affordability levels as pre-earthquake, many survivors were not able to recover their houses. Communication gaps also impacted recovery support, as residents who didn't follow the single track of housing recovery, including staying in temporary housing, for example, could miss out on information needed for support at later phases.

*Resourcefulness (Adaptive Capacities)*

As a wealthy country with strong governance mechanisms, Japan had strict compliance with building regulations pre- and post-earthquake. The Kobe Earthquake resulted in strong investments and commitments to various disaster risk reduction measures at local, regional, national, and international levels. Policies in Kobe set the direction for subsequent recovery projects, and processes of citizen participation built local connections and capacities for community-based recovery of government staff and other experts. However, not all communities or residents of Kobe actively participated in these processes, and not all survivors were supported by recovery programs.

## *Summary of Housing Recovery Outcomes in Kobe*

Residents had limited choices for significant aspects of housing recovery, regarding not only the design and allocation of disaster recovery public housing units, but also lots for private houses resulting from land readjustment. As many survivors were on their own, or left out of reconstruction projects, not all residents were able to recover their housing and lives after the earthquake (Koshiyama, 2011). As reconstruction targeted housing construction, there was limited support for economic and livelihood recovery. However, the government focus on providing disaster recovery public housing represents an approach of helping the most vulnerable survivors.

Kobe's *Machizukuri* Ordinance, which was established 15 years before the earthquake, shaped the culture of community involvement and planning in municipal recovery projects (stipulated by law) and fostered intensive community participation in recovery planning projects across the affected area. However, the focus of this participatory process was on neighborhood and community planning, with limited direct impact on housing reconstruction policy and programs, and therefore for the housing recovery paths of most survivors. Structures of community-planning councils still exist; however, 30 years later, many key members are aging, and finding people from younger generations to take on these responsibilities is a current challenge.

*Institutionalization of Housing Reconstruction over the Long Term*
As the first large urban disaster in the 50 years since World War II, housing recovery after the Kobe Earthquake required new policies and new approaches, which become the standard for subsequent modern disaster recovery in Japan. After realizing the negative impact of limitations on support for private housing recovery, these policies were modified later, although the underlaying logic is still consistent with Japanese approaches to post-disaster housing reconstruction today.

## Case 2: Housing Recovery After the 2011 Great East Japan Earthquake, Tsunami, and Nuclear Accident

On March 11, 2011, Japan was struck by a 9.0 magnitude earthquake, causing a massive tsunami that devastated 500 km of coastline in the northeast Tohoku region, and a meltdown at the Fukushima Daiichi Nuclear Power Plant. The official number of people who lost their lives was 19,765, with the bodies of an additional 2553 people who were never identified officially listed as "missing" (Reconstruction Agency, 2023), and 122,039 houses were completely damaged (Reconstruction Agency, 2023). Known as the Great East Japan Earthquake (GEJE) or 3.11 (after the date March 11), recovery needs were significant and varied after this massive, wide-area, and complex disaster. To manage recovery after this unprecedented disaster, the Japanese government created a new national Reconstruction Agency, and a new type of government subsidy, ***kofukin,*** funding a menu of recovery projects that municipalities could choose from to create their towns' recovery plans. Facing vast devastation, soon after the tsunami, the national government set the direction of recovery prioritizing risk reduction and moving away from areas at risk of future tsunami, including options for relocation of residential areas to high land and/or inland areas. Key housing recovery projects included Collective Relocation for Disaster Mitigation and provision of disaster recovery public housing. Including land readjustment projects, but mainly using collective relocation projects, lots for 18,226 houses were prepared in 393 project areas (Reconstruction Agency, 2023) and 29,654 units of public housing were planned.

The nuclear accident and meltdown at the Fukushima Daiichi Nuclear Power Plant released radioactive materials into the air, contaminating surrounding areas. Based on the degree of contamination, several levels of evacuation zones were created and revised multiple times. Facing long-term uncertainty and various complex factors, many nuclear evacuees were unable, or chose not, to return to affected areas, even after evacuation designations were lifted and they were allowed to return. Towards support for both the long-term recovery of the nuclear disaster-affected region, and the long-term needs of nuclear evacuees, many unresolved and unanswered questions persist. However, as the focus of this book is a close evaluation of long-term outcomes of housing reconstruction policy and programs, the case study in this chapter focuses not on the nuclear disaster-affected area but the tsunami-affected area where new housing has been built and occupied for 8 years.

The communities severely damaged by the tsunami stretch along the Tohoku (northeast) coast of Honshu, Japan's main and largest island, with the heaviest damage in Iwate, Miyagi, and Fukushima Prefectures. Geographically, this region includes the dramatic jagged fiord-like coastline (*rias* coast) to the north in Iwate and northern Miyagi Prefectures, dotted with fishing villages in small areas between the ocean and mountains. This geography funneled and amplified tsunami forces, washing away many communities, leaving limited land area available for rebuilding. Further south, in areas with the flat geography of the Sendai Plain in southern Miyagi and Fukushima Prefectures, tsunami waves caused destruction multiple kilometers inland, in some cases mitigated by elevated structures or highways, but leaving vast devastation in areas with no available high land for safer rebuilding.

Affected areas included towns and cities of various scales and economic activities, such as farming, fishing, industry, and tourism. While local culture and ways of life varied across disaster areas, affected communities included many places where traditional ways of life and work had prevailed before the tsunami. Coastal communities included a dense mix of residential, commercial, and industrial uses. However, even before the tsunami, this region was facing depopulation and aging even faster than this ongoing national demographic trend in Japan.

### *National Reconstruction Policies and Programs after the GEJE*

The Japanese government created a menu of nationally funded projects; each municipality could select projects to include in their town's recovery plan. Primary projects for housing reconstruction included Collective Relocation for Disaster Mitigation, land readjustment and/or land raising, and the creation and provision of disaster recovery public housing. The first part of collective relocation projects is the designation of former residential land as a tsunami hazard area, permanently forbidding new housing construction in that area and allowing owners to sell their former land to the government as part of the relocation project. In the second part of collective relocation projects, the government prepares sites and infrastructure for new residential lots in higher and/or inland areas. In Tohoku, residents participating in collective relocation projects had the option (depending on local government policy) to purchase or rent these lots, and rebuild their new homes.

As in Kobe, the provision of disaster recovery public housing was a large part of government support for disaster survivors who were not able or did not want to rebuild their own homes. As with all public housing in Japan, rents are government-subsidized and income-dependent. However, there were two major differences in disaster recovery public housing policies for disaster survivors after 2011. Unlike in Kobe, where eligibility for public housing was limited to low-income disaster survivors, all 3.11 survivors were eligible for disaster recovery public housing. However, after an initial period of low rents, later, income-based rents meant that for residents with income, rents could become higher than rent for apartments in the private housing market. Another difference was that along with multi-family apartment-style public housing units, like those built in Kobe, single-family detached public housing units were also constructed after subsequent disasters, including 3.11. Combined with collective relocation of communities, this policy allowed for the creation of both new housing lots (for residents to rebuild) and new public housing units (for residents to move into), co-located in or near the same relocation areas, relatively close to former neighborhoods, helping keep communities together. Single-family disaster recovery public housing was used extensively in smaller, traditional, and fishing coastal communities in tsunami-affected municipalities. Public housing built for disaster survivors in Shichigahama Town, the case study in this chapter, is one example; disaster recovery public housing in Shichigahama included many single-family units as well as some multi-family buildings.

*Case Study: Housing Recovery in Shichigahama Town*

**Pre-disaster Housing Context**

This case study focuses on one municipality in the tsunami-affected area, Shichigahama Town in Miyagi Prefecture. The smallest town in the prefecture, with an area of 13.19 square kilometers and a population of 17,683 (Shichigahama Town, 2023), the name Shichigahama means "Seven Beaches." The town is made up of multiple coastal communities, with traditional social structures of local government and community management still strongly in place. Local residents can trace their inhabitation over multiple generations; one interviewee was the 33rd generation of their family who had lived in the community. Less than 30 minutes by car from Sendai City, the largest city in the region, with open spaces and natural beaches popular for surfing and swimming, Shichigahama is also attractive for newcomers, commuters, and people who want to enjoy a relaxed lifestyle. Before 2011, many Shichigahama residents lived in single-family houses in coastal communities, including some older homes with traditional construction, passed down over multiple generations. Shichigahama also included some newer suburban-style housing developments in inland areas, home to more newcomers, outside of the tsunami-affected area.

As a very small municipality, where many residents and city staff know each other well and community structures go back many generations, Shichigahama has unique characteristics unlike other larger towns in the region; it is not representative of the housing recovery experiences of all tsunami-affected communities. However, it can also be considered as a microcosm through which to understand the different housing reconstruction projects implemented in tsunami-affected areas after 3.11, and the long-term outcomes of these projects for residents.

### Post-disaster Impacts and Housing Recovery Policies and Programs

On March 11, 2011, the magnitude 9 Great East Japan Earthquake caused strong shaking in Shichigahama at 2:46 pm. The first tsunami wave arrived at 3:51 pm, with inundation heights up to 12.1 meters, and tsunami waters inundating 4.8 square kilometers, or 36.4 percent of the town (Shichigahama Town, 2023). Ninety-four residents of Shichigahama lost their lives, as well as twelve people visiting from other towns. The bodies of two more people were never found, and three people passed away later from indirect impacts of the disaster. The number of houses damaged in Shichigahama was 3929, including 674 completely damaged houses, 237 severely damaged houses, 413 half-damaged houses, and 2605 with some damage (Shichigahama Town, 2023).

### Project Implementation

For tsunami-affected communities in Shichigahama, multiple collective relocation sites were created in highland areas, including both residential lots for private housing reconstruction and public housing. For communities with severe tsunami damage, mostly on the southern side of the town, areas were designated as hazardous zones (with residential construction forbidden), and new housing lots were prepared in highland areas using Collective Relocation for Disaster Mitigation Projects. In other areas with less damage from tsunami inundation, land readjustment projects (including land raising) were carried out for residents to rebuild or repair their homes in situ. Residents from these or other areas with less tsunami damage could also move into disaster recovery public housing built in highland areas.

Shichigahama's housing recovery projects targeted support for multiple communities, as follows.

- Matsugahama—13 lots for collective relocation, 32 public housing units
- Syoubutahama—30 lots for collective relocation, 100 public housing units, and land readjustment (4.1 hectares)
- Sasayama—128 lots for collective relocation
- Yoshidahama—9 lots for collective relocation, 6 public housing units
- Yogasakihama—14 lots for collective relocation, 24 public housing units, and two land readjustment projects (4.7 and 7.4 hectares, respectively)
- Hanabuchihama—40 public housing units, and land readjustment (9.8 hectares) (Shichigahama Town, 2018)

These various combinations of housing reconstruction projects reflect the numbers of households, scale of damage, and need for recovery support in each settlement. In total, Shichigahama Town supported the provision of 194 lots for collective relocation in 5 communities, 212 public housing units in 5 communities, and land readjustment projects in 4 districts, totaling 26 hectares (Shichigahama Town, 2018).

As a very small town, local government staff and local people knew each other very well. The town government took a personalized approach to housing recovery, helping each homeowner understand their options, related costs, and choices, to either rebuild in a new high land area as part of housing relocation projects or move into the provided disaster recovery public housing. Although a few residents moved away to other towns for personal reasons, almost all disaster survivors chose to stay and were able to regain a stable housing situation either by rebuilding in collective relocation settlements or moving into public housing (Fig. 4.3).

*Long-Term Outcomes Since the Disaster*

**Robustness and Diversity**

As in Kobe, all newly reconstructed housing in Shichigahama followed current building codes, including seismic requirements. Residents in both disaster recovery public housing and privately rebuilt housing reported confidence in their house's strength; some compared new houses favorably to pre-tsunami houses in this regard. Compared to several former residential areas with less solid ground, which magnified earthquake shaking and damage, new high land relocation settlements were built on stronger ground, and residents reported less shaking intensity and

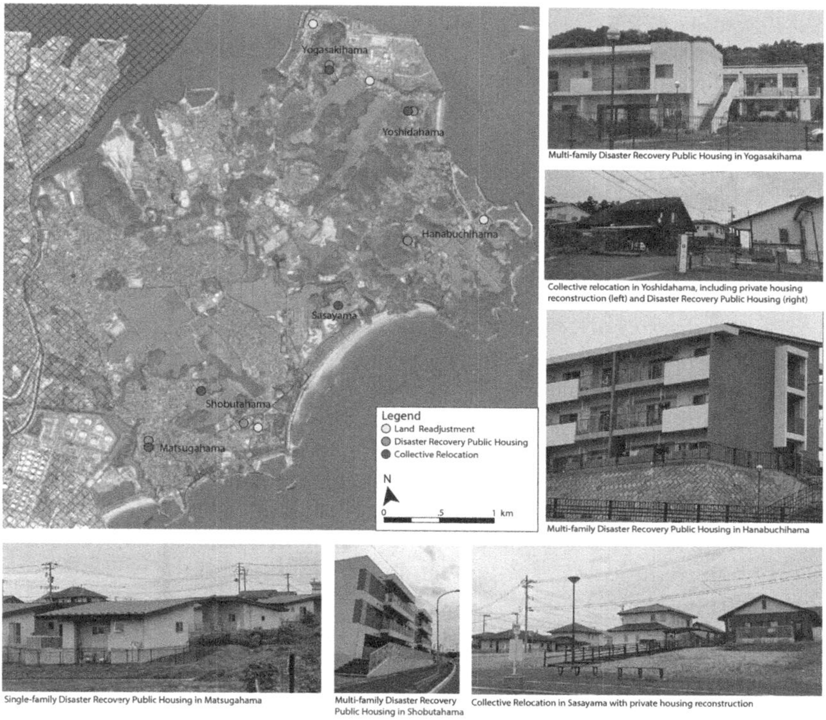


**Fig. 4.3** Housing reconstruction types, examples, and locations in Shichigahama after the 2011 Great East Japan Earthquake and Tsunami (Source: E. Maly, based on information from Shichigahama Town)

no damage during several large earthquakes in the past several years. After up to 8 years of inhabitation, no residents reported maintenance issues or concerns about related costs.

### Cultural Appropriateness and Adaptability

Shichigahama residents had lived in various kinds of houses pre-earthquake, including older traditional buildings and modern houses. Overall, they were satisfied with newly reconstructed houses, built to current housing standards, as warm, comfortable, and quiet, compared to older houses. Of course, residents who rebuilt their own homes could choose the design and layout. Residents who moved into apartment-style or single-family public housing units applied for certain areas or units, in some cases multiple times during the selection process, before being selected. Overall, residents reported that the houses were comfortable for their lifestyle, although in some cases smaller than former houses. Typical for contemporary Japanese housing culture, as public or private housing is fully built before inhabitation, residents rarely modified homes. In addition, as residents had only lived in their houses for up to 8 years, they reported no significant changes, either implemented or desired, in the structure or interior of houses.

**Redundancy and Integration**

Across the disaster-affected region, post-tsunami recovery incorporated disaster risk reduction, including a logic of redundancy in coastal protection against future tsunami damage, and coastal levees, or sea walls, were built to mitigate damage from smaller tsunamis, which have occurred every 30–40 years in this region over the last century. However, larger low-frequency high-impact tsunamis like the one that struck in 2011, predicted to occur once in 1000 years, are expected to overtop levees. With this understanding, the logic of multiple defenses in land use planning incorporates redundancy through land use planning and the relocation of housing to highland areas.

As in most tsunami-affected municipalities, new higher sea walls were constructed as part of recovery projects. In Shichigahama, the town made efforts to design attractive sea walls, including options for multi-use walking and cycling paths. Residents' opinions about the seawalls were mixed, as some residents felt they negatively impacted the environment and were not a sustainable choice, while others recognized their importance as a tsunami mitigation measure. With many pre-existing open areas, natural features, and beaches, many residents felt that too many parks had been created by Shichigahama's recovery planning (project requirements stipulated creating a certain amount of park or open space) that were unused, and conversely required maintenance. Consideration of green approaches to the actual reconstruction and building construction were very limited.

**Rights, Equity and Security**

As a small municipality with close connections between government staff and local residents, with very few exceptions of survivors who moved away for individual reasons, all survivors were able to recover their house in Shichigahama, either by rebuilding themselves on lots provided in collective relocation sites, or by moving into disaster recovery public housing. Most residents were able to move into their new houses by 2015, four years after the tsunami. As the reconstruction included relocation to highland areas, preparation of residential lots and infrastructure required additional time before housing construction could start. Considering this, residents felt that the required time was reasonable. One respondent interviewed in 2024, who had lived outside of the Shichigahama during the temporary housing phase, had less access to information about the planning process, which led to frustration because of not knowing the project timeline. However, he was able to return and move into a public housing apartment.

**Resourcefulness (Adaptive Capacities)**
As in Kobe before and after the 1995 earthquake, there was already strict enforcement of planning and building regulations before and after the 2011 tsunami. Facing a redevelopment project much larger than such a small town would usually see, the local town staff worked very hard over multiple years to plan and implement recovery projects. They were also supported by officials dispatched from other regions in Japan, including former retired government workers; this system, developed after past disasters, was used across the tsunami-affected area to support the limited human resources of small towns.

Local government dealt with new types of projects but worked closely with local residents to help them understand their options and recover their housing. Although most tsunami-affected municipalities also had dedicated municipal staff working for town and community recovery, Shichighama's approach can be understood as exceptionally sensitive to residents' needs, with one-on-one housing recovery advice provided to each household. Also reflecting an increase in disaster awareness and related policies across the region and the country after the GEJE, there have also been many developments and advancements in community-based disaster preparedness in Shichigahama since the tsunami until today, including community-based disaster risk reduction activities such as evacuation planning and drills.

## *Comparative Analysis of Two Cases from Japan*

After providing an overview of similarities, differences, and changes in housing reconstruction policy and project implementation learned from these two cases from Japan, this section compares the long-term outcomes in the context of resilience characteristics, shown in Table 4.1. Finally, drawing from the two cases, key factors that contribute to catalyzing longer-term resilient outcomes in Japan are analyzed.

**Table 4.1** Comparing resilience characteristics as long-term outcomes of two housing reconstruction projects in Japan (●●● = strongly agree; ◎ = agree; ± = neutral) (Source: E. Maly)

| *Resilience characteristics* | *Resilience outcomes in housing systems over the long term* | *Description of long-term resilience outcomes in housing, communities and institutions* | *Kobe 1995 EQ (urban)* | *Shichigahama 2011 GEJE tsunami (small town)* |
|---|---|---|---|---|
| Robustness and diversity (buffer capacities) | Planning and building standards | Well located for reduced exposure | ◎ | ●●● |
| | | Building standards stronger/revised | ●●● | ●●● |
| | Structural safety | Structural hazard-safety features explicitly integrated in housing | ●●● | ●●● |
| | | House withstood another disruption | ●●● | ●●● |
| | | Good construction quality | ●●● | ●●● |
| | | Choice of materials and construction technologies | ●●● | ●●● |
| | | Traditional/local construction labor incorporated | ± | ± |
| | | Replication of improved construction systems | ± | ± |
| Cultural appropriateness and adaptability | Housing design | Design with diverse housing typologies | ◎ | ●●● |
| | | Culturally responsive spatial layout | ◎ | ◎ |
| | | Thermally comfortable design | ●●● | ●●● |
| | | Design flexibility for extension or adaptation | ◎ | ◎ |
| | | Occupancy in houses | ◎ | ◎ |
| Redundancy and integration | Settlement or city development | Services integrated with housing | ●●● | ●●● |
| | | Infrastructure integrated with settlement | ●●● | ●●● |
| | | Social amenities integrated in settlement | ± | ◎ |
| | | Livelihood proximity/considerations | ◎ | ◎ |
| | | Redundancy integrated in housing or settlement | ± | ◎ |

| | | | | |
|---|---|---|---|---|
| Rights, equity, and security | Equity, inclusion | Recovery for all residents | ◎ | ●●● |
| | Dignity, sense of ownership | Participatory process | ◎ | ●●● |
| | Affordability | Low cost, affordable, and ease of maintenance | ◎ | ◎ |
| | Security | Better security of land, housing tenure | ± | ± |
| Resourcefulness (adaptive capacities) | Institutional capacities: recovery program, assistance, agility, rapidity, timely, yet strategic | Enforcement of planning and building regulations | ●●● | ●●● |
| | | Improved governance and institutional capacity | ◎ | ◎ |
| | | Proactive investment in prevention strategies | ●●● | ●●● |
| | Social capacities | Changed social mindset, norm, and improved or preserved social bond | ●●● | ◎ |
| | Individual capacities | Improved household awareness, security, and reduced aid dependency. | ◎ | ◎ |
| | | Better quality of life and political voice | ◎ | ◎ |

## Similarities, Differences, and Changes in Housing Reconstruction Projects over Time

In a comparison of housing recovery policy for households affected by the 1995 Great Hanshin Awaji Earthquake and the 2011 Great East Japan Earthquake and Tsunami, several differences are found due to unique aspects of the respective disasters, including the nature of post-earthquake vs post-tsunami recovery, and respective sizes and scales of disaster damage and recovery. There are strong similarities between the two cases, as Japan's national government plays a leading role in disaster recovery, leading to consistent recovery policies across cases, although with some changes over time. 30 years and 14 years later, respectively, long-term outcomes after these two disasters in Japan share commonalities, as well as reflecting impacts of several key reconstruction policy differences.

### *Robustness and Diversity*

Due to Japan's high building standards and construction compliance, high levels of technical robustness at the housing scale are consistent in Japan, both before and after the 1995 earthquake and 2011 tsunami. As a result, housing reconstruction improved robustness of new houses of residents whose older houses were damaged in either disaster, and high levels of robustness were retained over long-term housing reconstruction outcomes for both cases. Housing reconstruction at the settlement scale also showed increased robustness in both Japanese cases, although Kobe's post-earthquake urban recovery made use of land readjustment within neighborhoods, while Shichigahama's post-tsunami recovery used relocation projects to move residents away from coastal areas to higher elevations, reducing future tsunami exposure.

### *Cultural Appropriateness and Adaptability*

In terms of long-term social-cultural outcomes, there were significant changes and improvements in housing reconstruction in Shichigahama after the 2011 tsunami, compared to housing recovery in Kobe after the 1995 earthquake. Housing recovery in both cases focused on providing fully built disaster recovery public housing. However, after the 2011 GEJE, single-family detached public housing units were also provided, along with multi-family public housing. The increased range of housing types allowed greater continuity from pre-disaster living environments and local culture for small coastal communities. Shichigahama, where single-family disaster recovery public housing units were built along with other lots for private housing reconstruction in high land relocation areas, is one such example. Drawing from lessons after the 1995 earthquake, the importance of keeping communities together during the recovery process was well understood and supported by the collective relocation of communities to nearby high land areas after 3.11, including both lots for self-built recovery, together with the construction of single-family disaster recovery public housing units.

### *Redundancy and Integration*

Both cases of Kobe and Shichigahama demonstrated positive long-term outcomes in terms of integrated settlement design and land use planning, and relocation that incorporated redundancy for risk reduction. However, land use planning strategies differed in both cases, with post-earthquake reconstruction in an urban area using land use readjustment and redevelopment projects in Kobe after 1995, and relocation projects after the 2011 tsunami moving residents away from areas at risk of future tsunami damage.

### *Rights, Equity, and Security*

Changes in policy between 1995 and 2011 also represented an expansion of housing recovery support that improved equity in long-term outcomes for survivors. One major policy change was the expansion of support for residents to rebuild their own houses available after 2011, which had not been available in 1995. After the GEJE, collective relocation projects, including the preparation of new residential lots, represented more choices for affected households. Due to the differences in earthquake and tsunami risk and damage, the large number of housing relocation projects after the tsunami is not only different from Kobe's post-earthquake recovery, but unprecedented in Japan as well as on a global level.

### *Resourcefulness (Adaptive Capacity)*

Both cases included aspects that developed the capacity of communities, organizations, and the government in the recovery process, leading to positive long-term outcomes. At the community level, planning processes with residents' participation had been newly developed in Kobe in the decade before the earthquake through the *machizukuri* ordinance, which played a key role in post-earthquake recovery planning, leading to strong community capacity over the last 30 years, along with a legacy of expert and government capacity. Residents in Shichigahama also participated in recovery planning, where strong traditional community structures contributed to smooth implementation of reconstruction projects. Close relationships between local government staff and local residents led to a careful and detailed approach, advising each household on their options for rebuilding and helping them make decisions. As in this case, small communities where everyone knows each other means that communication is in some ways easier; however, there is also the risk that dominant voices may overpower less vocal ones.

There were also positive long-term outcomes for building government capacity after both disasters. As a metropolitan area, Kobe City had many staff working in the housing department with prior experience in urban development. However, in 1995, there were no precedents for large-scale post-disaster reconstruction, and post-disaster recovery planning was a new endeavor. Although the Kobe Earthquake came as a surprise to many, it led to the broad development of community-based as well as government-driven disaster risk reduction activities and initiatives in Japan.

With the large scale of damage, awareness of tsunami risk, and need for recovery and reconstruction over a large area, some aspects of reconstruction programs after 3.11 also represent new and unprecedented approaches in Japan, including the creation of the national reconstruction agency and the *kofukin* subsidy budget through which the national government funded almost 100% of reconstruction projects. The commitment to future tsunami risk mitigation led to large-scale relocation in projects implemented across the tsunami-affected area. However, there is a risk of future challenges for the sustainability of these communities, created in primarily single-use neighborhoods, with aging residents.

## Key Factors with the Potential to Catalyze Resilient Long-Term Outcomes

From a comparison of the implementation and long-term outcomes of housing reconstruction projects from the two cases in Japan, the following key factors can be understood as having shaped the Japanese recovery context and impacted community resilience, and suggest proposed points of comparison with other countries' experiences.

### *Leading Role of Government*

Although Japan has a seen a growing role of volunteers, non-profit and civil society organizations involved in supporting disaster survivors and community recovery, housing reconstruction support was and still is almost completely government-driven. The national government provides most reconstruction funding, in line with government policies and recovery projects. Although the limited role of civil society organizations has expanded since 1995 and became more active after 2011, there is still an untapped potential for a greater role of CSOs to provide additional and more varied support to survivors, and contribute to additional equity in response, recovery, and long-term outcomes.

### *Strong Governance and Compliance*

Strong governance and compliance with regulations in Japan also shaped recovery in both cases, from the high standards for strong building construction and the ability to implement rigorous land use planning and relocation projects, leading to long-term recovery with robustness at both housing and settlement scales. With strong legal systems, building codes, and compliance, construction in Japan is built to high standards. Older buildings built before the 1980s may have been weaker and suffered more disaster damage, but unlike other countries where there is a risk of newly reconstructed buildings that don't meet required codes or standards, new construction in Japan is reliably built to code, whether single-family houses built by homeowners using private companies, public or private housing redevelopment, or government-built single- or multi-family housing structures.

### *Land Use and Settlement Planning*

Recovery after earthquake and tsunami damage included different aspects related to land use planning for mitigation. However, in both cases, government-driven housing reconstruction included land use planning for risk reduction—land readjustment in Kobe and collective residential relocation of tsunami-affected communities in Tohoku. In Kobe, where many of the most-affected areas after the earthquake were densely built low-rise neighborhoods that suffered from post-earthquake fires, land readjustment and redevelopment projects created wider streets and more open areas. As a modern metropolis, much of Kobe's recovery also included residents rebuilding their own houses on site. However, in some (especially older) areas of the city, there was also improvement in settlement-scale infrastructure.

With the large-scale use of collective relocation projects after the tsunami, housing reconstruction in the Tohoku region includes a significant alteration of townscapes, landscapes, and community location. Whereas the many collective relocation sites were built at higher elevation levels, with the goal of reducing future tsunami risk, other challenges may emerge as these kinds of recovery projects replace former dense and multi-use traditional towns with separated uses and increased distance from the sea. Housing recovery projects in Kobe and Tohoku did not include significant consideration of sustainable development, building materials, or environmental issues for planning or construction.

### *More Housing Reconstruction Options for Residents*

In terms of direct support for housing reconstruction, there are strong similarities and continuity between the two cases, including the shared focus on provision of in-kind subsidized rental housing in the form of disaster recovery public housing. However, housing reconstruction policy to support homeowners in rebuilding their own houses was revised significantly between the two disasters, with more support available to homeowners in 2011. In addition, collective relocation projects after the 2011 tsunami combined prepared residential lots on which homeowners could rebuild, with smaller-scale single-family public housing.

### *Capacity Building of Community and Government for Disaster Risk Reduction and Reconstruction*

In terms of government capacity, as a large city, Kobe had a much larger capacity in terms of human resources and staff in the city's housing or planning departments that could be deployed to support disaster recovery. In smaller and more rural towns in tsunami-affected municipalities in Tohoku, there were fewer staff members, and even fewer with experience in urban redevelopment projects at the large scales of recovery projects, rare for this region in non-disaster times. Policies and programs provided by the national government attempted to address this issue, along with the massive scale of the disaster and the intense need for recovery support across multiple prefectures and municipalities at the same time, through the creation of standardized options for recovery planning.

In Kobe, one of the earliest cities in Japan to pass a *machizukuri* ordinance, there was already existing capacity to carry out participatory community planning as part of recovery projects. In addition, some communities facing recovery planning had already started community-planning processes before the earthquake. Official recovery projects created a hierarchy of target areas, with the "black" (most severely damaged) areas targeted for the majority of recovery projects, "gray" areas eligible for some assistance, and almost no recovery support for the relatively undamaged "white" areas. However, there may have been some differences between active communities from or including "white" areas, and those without effective leadership.

In traditional communities like Shichigahama, strong community ties and familiarity between local town staff, community leaders, and residents contributed to inclusive recovery processes. In areas where traditional community structures and hierarchy were still in place, these structures could make for smoother implementation of recovery projects, and with close multi-generation connections and familiarity between local government staff, community leaders, and residents, there were few residents left out of housing recovery, as support programs could be provided and customized for each resident.

In both Kobe and Tohoku, residents who actively participated in recovery planning processes contributed to building community capacity. However, as in both regions community leaders are aging and face challenges related to limited numbers of younger people, there are concerns about how this capacity will be passed on to future generations.

## Conclusion

Housing recovery policies in Japan have evolved since 1995, including through the recovery processes of other disasters that have occurred in the meantime. The scale and complexity of the damage and recovery needed after the 2011 Great East Japan Earthquake were unprecedented in Japan, and post-3.11 recovery included some new aspects introduced in response to this challenge. At the same time, the fundamental approach of modern post-World War II post-disaster housing recovery in Japan remained relatively constant, as did the government's main role as the driver of reconstruction.

Japan already had high standards of building construction including seismic codes and modern infrastructure before both disasters. New housing built as post-disaster reconstruction represents the same or improved housing quality for residents, especially if their pre-disaster housing was relatively older. After 2011, new housing construction built to current standards had improved insulation and was therefore warmer and quieter.

Although relocation to higher land areas after the GEJE reduced the risk of exposure to future tsunamis, relocated communities may face future challenges for sustainable neighborhoods, as new uses of these areas include only housing, and the population is aging as well as shrinking. Housing recovery projects in Kobe and Tohoku did not include any significant consideration of sustainable development, building materials, or consideration of environmental issues for planning or construction.

As a country with significant hazard risk, there have been several more large disasters in Japan since 3.11, including the 2016 Kumamoto Earthquake and 2024 Noto Earthquake, and there is a high probability of a large earthquake striking Tokyo, and a Nankai Trough earthquake that would generate a devastating large tsunami in southwest Japan. Precedents for Japanese housing recovery policy and programs are well established and relatively consistent, with some incremental improvements between the two case studies in this chapter. However, looking forward to future disaster reconstruction, increasing involvement of CSOs, along with participation and more varied choices for survivors, could transform recovery processes and build additional resilience capacity.

## References

Be Kobe. (2015). *What a Researcher in the Field of Disaster Risk Reduction Learned When he Became a Disaster Victim*. Retrieved April 12, 2024, from https://bekobe.smartkobe-portal.com/interview/en/2016/01/2324/index.html

Cabinet Office. (2014). *Disaster Management in Japan*. Retrieved April 12, 2024, from https://www.bousai.go.jp/1info/pdf/saigaipanf_e.pdf

Edgington, D. W. (2010). *Reconstructing Kobe: The Geography of Crisis and Opportunity*. University of British Columbia Press.

Hirayama, Y. (2000). Collapse and Reconstruction: Housing Recovery Policy in Kobe after the Hanshin Great Earthquake, Housing Studies. 15 111–128.

Koshiyama, K. (2011). Comparison of International and Domestic Methods of Providing Housing after Disasters. *Journal of Disaster Research, 6*(2), 230–235.

Maki, N. (2007). How Can Public Sector Support Recovery of Privately Owned Individual Housing after Natural Disaster? Possibility of Setting Housing Recovery Grant in Japan. In *2nd International Conference on Urban Disaster Reduction, 27–29 November 2007.*

Maly, E. (2025). Japan's Disaster Risk Reduction Diplomacy and the Sendai Framework for Disaster Risk Reduction. In P. Cavaliere & J. Otani (Eds.), *Handbook of Disaster Studies in Japan.* MHM Limited (In press for 2025). Retrieved from https://www.mhmjapandocuments.com/handbook-series

Murosaki, Y. (2007). Lessons on Reconstruction Strategies from the Great Hanshin-Awaji Earthquake. *Journal of Disaster Research, 2*(5), 330–334.

Olshansky, R. B., Johnson, L. A., & Topping, K. (with Murosaki, Y., Ohnishi, K., Koura, H., & Kobayashi, I.). (2005). *Opportunity in Chaos: Rebuilding After the 1994 Northridge and 1995 Kobe Earthquakes.* University of Illinois, Department of Urban and Regional Planning. Retrieved from http://www.urban.illinois.edu/faculty/olshansky/chaos/chaos.html

Satoh, S. (2019). Evolution and Methodology of Japanese Machizukuri for the Improvement of Living Environments. *Japan Architectural Review, 2,* 127–142.

Shichigahama Town. (2018). *Shichigaham Reconstructuion Overview.* Retrieved December 4, 2024, from https://www.shichigahama.com/town2/documents/gaikyo_e.pdf

Shichigahama Town, Recovery Update. 2023

CHAPTER 5

# Cross-Cultural Insights from Long-Term Analysis of Post-Disaster Housing Reconstruction to Resilient Recovery Outcomes

**Abstract** This concluding chapter extends the conversation to a cross-cultural synthesis, distilling similarities, differences, and observable shifts in policies and practice that bear on enhancing long-term resilience. Critically analyzing suggestions from empirical evidence about effective/catalytic strategies in resilient recovery, we conclude by putting forward a few recommendations that emerged. We hope to inform future recovery interventions, domestically, regionally within the Asia-Pacific, and internationally, towards the resilience imperative of our times. Recommendations include five key factors for successful housing recovery: (1) Inclusive governance and coordination mechanisms; (2) Government–civil society organization (CSO) partnerships for successful implementation of owner-driven housing reconstruction (ODHR) programs (in India and Thailand); (3) Proactive revision of land use planning and building standards, for hazard exposure reduction and hazard mitigation; (4) Timely, equitable, inclusive, and appropriate resource allocation and distribution; and (5) Strengthening capacities at all levels of government, the construction sector, and communities. These factors were essential in fostering resilient outcomes and building capacities for hazard mitigation at multiple scales—individual, community, and government/institutional levels, as well as various sectors (planning, design, and construction). The study also identified factors that hindered resilience outcomes in certain cases, such as hazard type (scale and element of surprise) as it affected provision or lack



M. Vahanvati et al., *Resilient Recovery from Disasters*,
https://doi.org/10.1007/978-981-97-8047-1_5

thereof of financial assistance for recovery; shortfalls of predominantly market-driven recovery programs; and/or governments' top-down coordination. The most significant factor, with implications across all the five recommendations, was the importance of sustaining and nurturing capacities built during housing reconstruction and linking these to ongoing development activities, to truly transition towards risk-informed resilient development pathways. While each country and disaster situation are unique, requiring contextually tailored responses, some lessons can be generalized based on learnings over decades, and we hope they will inform future interventions to avoid repeating the same mistakes.

**Keywords** Asia • Cross-cultural • Post disaster • Housing reconstruction • Disaster resilience • Long-term recovery • Resilient recovery • Recovery • Build Back Better/Build Back Safer • Adaptive capacities • Capacities • Robustness • Redundancy • Rights • Systemic changes

## Comparative Analysis of Longer-Term Resilience Outcomes in India, Thailand, and Japan

Several key similarities and differences emerged from cross-country comparison of the six case studies in the three Asian countries—India, Thailand, and Japan, over a long period of time (14 to 30 years ago), as shown in Table 5.1. Japan's recovery was highly centralized, extensively financed by the government, achieved the highest technical robustness of housing construction and settlements, and included aspects that supported equitable recovery. However, this was not achieved for all survivors. India, on the other hand, had a highly collaborative governance approach (i.e. ODHR policy later transforming into agile policy), innovative recovery coordination (through *Setu Kendras*), active roles of CSOs, and funding from multiple sources, in addition to a major loan from the World Bank and inclusion of local traditional and sustainable construction technologies in building standards. However, India faced and continues to face challenges in complying with building standards and sustaining capacity building. Thailand is positioned somewhere between India and Japan, with centralized decision making for recovery policy and governance and funding from multiple sources, resulting in diverse housing outcomes and community participatory aspects.

**Table 5.1** Comparing resilience characteristics as long-term outcomes of post-disaster housing reconstruction projects in India, Thailand, and Japan. ●●● = strongly agree; ◎ = agree; ± = neutral; ⊠ = disagree; ⊠⊠⊠ = strongly disagree. (Source: Authors)

| | *Resilience outcomes in housing systems over the long term* | *Description of long-term resilience outcomes in housing, communities, and institutions* | *India case studies* | | *Thailand case studies* | | *Japan case studies* | |
|---|---|---|---|---|---|---|---|---|
| | | | *GIDC Bhuj 2001 Gujarat EQ (urban)* | *Manganj 2008 Bihar flood (rural)* | *Nam Khem 2004 tsunami (rural)* | *Ayutthaya, 2011 flood (urban)* | *Kobe 1995 EQ (urban)* | *Shichiga-hama 2011 tsunami (small town)* |
| Robustness and diversity | Planning and building standards | Well located for reduced exposure | ⊠⊠⊠ | ⊠⊠⊠ | ◎ | ⊠ | ◎ | ●●● |
| | | Building standards stronger/revised | ●●● | ●●● | ⊠ | ⊠ | ●●● | ●●● |
| | Structural safety | Structural hazard-safety features explicitly integrated in housing | ●●● | ◎ | ◎ | ◎ | ●●● | ●●● |
| | | House withstood another disruption | ●●● | ●●● | ⊠ | ⊠ | ●●● | ●●● |
| | | Good construction quality | ●●● | ◎ | ◎ | ◎ | ●●● | ●●● |
| | | Choice of materials and construction technologies | ●●● | ●●● | ◎ | ⊠ | ●●● | ●●● |
| | | Traditional/ local construction labor incorporated | ●●● | ●●● | ●●● | ●●● | ± | ± |
| | | Replication of improved construction systems | ⊠ | ◎ | ⊠ | ⊠ | ± | ± |

*(continued)*

**Table 5.1 (continued)**

| | *Resilience outcomes in housing system in the long term* | *Description of long-term resilience outcomes in housing, communities, and institutions* | *Indian case studies* | | *Thailand case studies* | | *Japan case studies* | |
|---|---|---|---|---|---|---|---|---|
| | | | *GIDC, 2001 Gujarat EQ (urban)* | *Manganj 2008 Bihar flood (rural)* | *Nam Khem 2004 tsunami (rural)* | *Ayutthaya, 2011, flood (urban)* | *Kobe 1995 EQ (urban)* | *Shichiga-hama 2011 tsunami (small town)* |
| Cultural appropriateness and adaptability | Housing design | Design with diverse housing typologies | ◎ | ●●● | ●●● | ± | ◎ | ●●● |
| | | Culturally responsive spatial layout | ●●● | ●●● | ●●● | ◎ | ◎ | ◎ |
| | | Thermally comfortable design | ●●● | ●●● | ± | ⊠ | ●●● | ●●● |
| | | Design flexibility for extension or adaptation | ◎ | ●●● | ●●● | ± | ◎ | ◎ |
| | | Occupancy in houses | ⊠ | ●●● | ◎ | ◎ | ◎ | ◎ |
| Redundancy and integration | Settlement or city development | Services integrated with housing | ●●● | ◎ | ●●● | ●●● | ●●● | ●●● |
| | | Infrastructure integrated with settlement | ⊠ | ◎ | ●●● | ●●● | ●●● | ●●● |
| | | Social amenities integrated with settlement | ●●● | ± | ●●● | ●●● | ± | ◎ |
| | | Livelihood proximity/ considerations | ⊠⊠⊠ | ± | ◎ | ◎ | ◎ | ◎ |
| | | Redundancy integrated in housing or settlement | ◎ | ◎ | ◎ | ◎ | ± | ◎ |

| | | | | | | | | |
|---|---|---|---|---|---|---|---|---|
| Rights, equity and security | Equity, inclusion, dignity and security | Recovery for all residents | ± | ± | ◎ | ± | ◎ | ●●● |
| | | Participatory process | ●●● | ●●● | ●●● | ◎ | ◎ | ●●● |
| | | Low cost, affordable, and ease of maintenance | ◎ | ●●● | ●●● | ◎ | ◎ | ◎ |
| | | Better security of land, housing tenure | ± | ± | ●●● | ●●● | ± | ± |
| Resourcefulness (adaptive capacities) | Institutional capacities: recovery program, assistance, agility, rapidity, timely, yet strategic | Enforcement of planning and building regulations | ☒☒☒ | ☒☒☒ | ◎ | ◎ | ●●● | ●●● |
| | | Improved governance and institutional capacity | ◎ | ± | ± | ± | ◎ | ◎ |
| | | Proactive investment in prevention strategies | ◎ | ± | ± | ± | ●●● | ●●● |
| | Social capacities | Changed social mindset, norm, and improved or preserved social bond | ± | ◎ | ●●● | ●●● | ●●● | ◎ |
| | Individual capacities | Improved household awareness, security, and reduced aid dependency | ◎ | ◎ | ●●● | ◎ | ◎ | ◎ |
| | | Better quality of life and political voice | ◎ | ± | ◎ | ◎ | ◎ | ◎ |

### *Robustness and Diversity (Buffer Capacities)*

Robustness relates to resilience of hard/physical assets. Of the three countries, Japan has high levels of technical robustness, followed by India, and then Thailand. Robustness at housing as well as settlement scales is high in Japan, due to the country's strong building standards and enforced construction practices following detailed and strict building codes. Not limited to reconstruction, Japan demonstrates almost universal compliance with building standards, which means that most housing damage from earthquakes was to older buildings built before the current code (introduced in 1981), and any new buildings constructed before or after either disaster would be built to current standards. Conversely, as standards were already high, reconstruction policy had a limited impact on changing standards.

Following Japan, India's case study settlements—both the urban relocation site of GIDC in Gujarat, as well as rural in-situ reconstruction in Bihar, attained high levels of technical robustness at the housing scale. Following India, in Thailand's case study settlements, such improvements were only seen in certain projects post-tsunami, and for only a small fraction of post-flood reconstruction projects. Governments in Japan (in 1981, years before the case study disasters) and India revised their seismic regulations and building codes. India was the only country that legalized and integrated traditional construction technologies in building standards, including stabilized earth and stone construction after the 2001 Gujarat Earthquake and bamboo construction after the 2008 Kosi River floods in Bihar. Although green and bio-based technologies were legalized and encouraged by CSOs, their acceptance by residents in both case studies in India (Bhuj and Bihar) declined over the mid-term (~7 years after the case study disasters), as discussed by Vahanvati (2015), and almost disappeared over the longer term. In the long term, most houses in Bihar's rural towns have been replaced by brick and RCC construction, demonstrating a gradual shift towards more industrial materials. In Thailand's Namkhem communities, houses built by construction professionals, paid for by donors or with the support of Community Organization Development Institute (CODI), were of much higher construction quality.

Unlike Japan, India and Thailand experienced challenges sustaining the training of construction labor, enforcement of compliance with building standards, and hence robust and high-quality construction. Nonetheless, in case studies from all three countries, housing reconstructed after disasters, despite varying robustness, has stood the test of time and withstood subsequent disasters, suggesting some improvements, if not radical or systemic changes.

### *Cultural Appropriateness and Adaptability*

From the social and cultural perspective, India and Thailand attained high levels of culturally appropriate housing and high levels of resident satisfaction. With housing design, both in urban and rural areas, fairly similar to pre-disaster housing designs, overall, all cases showed relatively high degrees of cultural appropriateness. In cases where houses were designed by their occupants, such as in India's owner-driven reconstruction in Gujarat and Bihar, or in Thailand's community-driven resettlement in Namkhem, and to some extent in Japan (when homeowners designed and rebuilt their own homes), resident satisfaction was high, as their houses met their needs.

Additionally, in case studies in all three countries, residents were highly satisfied with stronger houses, improved thermal comfort, and reduced maintenance needs. In Thailand, this varied based on housing provision type post-tsunami in Namkhem, and as post-flood support in Ayutthaya was insufficient to support reconstruction. While the design of houses remained culturally and climatically responsive, in all case studies, material and construction technology aspects of reconstructed houses have incrementally changed from traditional to more modern ones (particularly in India and Thailand). For example, in India's Maanganj village of Bihar, since the 2008 Kosi River floods, traditional bio-based (bamboo and thatch) houses have incrementally been replaced by brick and RCC-constructed houses. In both cases in Japan, disasters damaged some houses with traditional construction, and almost all reconstructed houses used modern construction and materials. In Thailand, housing culture in Ayutthaya had already moved away from traditional houses before the 2011 floods. Houses rebuilt in Namkhem, Thailand after the 2004 tsunami with the support of CODI were much stronger and comfortable than before, and some communities have also been able to extend and improve their housing over time.

Social connection and settlement-scale spatial layout were often compromised to attain higher robustness, as evident in both Japan case studies, Thailand's Namkhem case post-tsunami, and India's GIDC urban relocation case post-earthquake. Japan's new, modern, and stronger housing has not always responded to its residents' cultural needs. In Japan, a large share of reconstructed housing after both the 1995 earthquake and the 2011 earthquake and tsunami replaced older wooden, low-rise houses with stronger and modern reinforced concrete buildings. In 1995, the newly built, multi-family public housing apartments were government-provided, with no resident involvement in the design process or choice of their specific units. Such sudden changes in the living environment and living culture, away from neighbors or former communities, and to living in isolation in tall apartment buildings had severe impacts on the mental health, social connection, and well-being of residents, especially the elderly. Learning from experiences after the 1995 earthquake regarding residents' need for social connection, reconstruction after the 2011 earthquake and tsunami provided a variety of public housing, including multi-family apartment-style housing and single-family detached housing, especially in smaller communities, to keep residents closer to former neighborhoods and neighbors. Similarly, in Thailand, the default recovery housing provided by the government and constructed by the military in Namkhem was a small and standard-sized design, which was not satisfactory for residents. In the same vein, in India, after the 2001 Gujarat Earthquake, the renters who were relocated from the old city to the GIDC relocation site were not given an option of choosing their neighbors, resulting in mixed communities and loss of social bonding. In the long-term, residents are unhappy with the social mix, and a lot of residents have moved out of GIDC to live closer to their community. GIDC's case also highlights India's deep-rooted challenges associated with societal stratification along caste and religious lines.

Only a few case studies struck a balance between houses' technical robustness and cultural and livelihood needs, through a participatory process. In one example, Thai communities whose livelihood relied on crab fishing and living above water were assisted by CODI to rebuild their houses over water but on reinforced concrete pillars. In another example, CODI supported indigenous communities to reconstruct houses using timber but with a raised second floor. Similarly, in India's Bihar after the 2008 Kosi River floods, the community's need to live close to fertile agricultural land in flood-prone areas was respected, and they were allowed to rebuild their houses in situ on reinforced concrete structures with deep pile footings and raised floors with attic space.

### *Redundancy and Integration (Buffer Capacities)*

Japan has been proactively championing risk reduction activities through future hazard risk-informed land use planning regulations after both the 1995 earthquake and 2011 earthquake and tsunami. Land readjustment projects in Kobe after the 1995 earthquake resulted in wider streets and open areas. Of the three countries, only Japan's national government reconstruction policy actively engaged in municipal-scale development planning after 2011, revising land uses in tsunami-affected areas and proactively relocating residents away from hazardous areas.

The only other example that included risk reduction through comprehensive land use planning changes was after the 2001 Gujarat Earthquake—Bhuj walled city's urban renewal project. Land readjustment of Bhuj's walled city resulted in wider streets, open spaces and relocation of many residents, managed by the state and local government. The GIDC relocation site also included comprehensive settlement planning by local CSOs (see Fig. 2.3). However, the urban renewal plan and land readjustment did not reduce Bhuj's communities' exposure to earthquakes because a vast portion of the town is earthquake prone. In Thailand after the tsunami, the risk zone was studied and publicized, and some donor-supported housing relocation settlements moved some residents away from the ocean, although housing reconstruction did not require relocation out of high-risk zones. However, some new donor-supported housing relocation projects moved residents to safer areas. No land use planning or settlement-scale changes were evident in India's rural areas of Gujarat after the 2001 earthquake or Bihar after the 2008 Kosi River floods, or in Thailand's Namkhem after the 2004 tsunami and Ayutthaya after the 2011 floods. Reducing risk exposure of communities from future hazards requires risk mapping data, availability of low-risk land, and the ability to implement relocation of residents, which continues to be a major challenge in India and Thailand.

Redundancies for mitigation can be implemented at multiple scales—housing, settlements, and/or services—to allow individual measures or structures to fail without a major impact on housing or communities. Such redundancies were incorporated in all countries, but at varied levels. At the settlement scale, redundancies were incorporated in case studies in all three countries. In Thailand, restored mangrove forests contributed to future tsunami mitigation. In Japan, multiple layers of tsunami defenses included the construction of levees, combined with land use planning and relocation of housing to higher ground. In India, redundancies were evident in services at settlement scale in Bhuj city, where green water sewage recycling and garbage conversion to vermicompost were integrated in GIDC's settlements for drought proofing and guaranteeing water and fertile soil even when the town's water supply fails. In India's Bihar, redundancies in housing included break-away ground floor walls that can give way during floods with the surviving house sitting on a raised structure supported by deep pile footings. This form of housing is also termed "wet architecture."

### *Rights and Equity*

Japan's housing reconstruction support can be understood as a social welfare approach, with the focus after both the 1995 earthquake and 2011 tsunami on providing disaster recovery public housing to low-income communities. Although according to policy, housing recovery support was available to all survivors who needed it, the actual support provided did not match all residents' needs. While the overall number of housing units damaged in the Kobe Earthquake were replaced, housing in certain areas was not replaced, and not every resident living in former neighborhoods regained a replacement house. In addition, with a strong single-track housing recovery process (temporary housing, followed by permanent housing), survivors who did not move into temporary housing were not always able to receive public housing later, and were at risk of not receiving all the information needed to access support, resulting in some residents missing out on housing support. With exclusive focus on public housing and almost no support for private housing repair or reconstruction, another group of survivors was left out of support policies—those who might have been able to rebuild their housing with minimal support but could not do so on their own. Some very low-income survivors, who were provided with subsidized rental public housing, also struggled with the required additional costs for moving in and utilities.

After the 2011 earthquake and tsunami, additional efforts were in place to provide outreach and recovery support for survivors throughout the recovery process. One example of such outreach was disaster case managers, who provide one-on-one help to survivors to access available support; however, these programs were only available in specific, limited areas. Despite the policy approaches of social welfare and equal support for all survivors, there were people still left out of the available recovery programs for various reasons. In contrast, there were also examples, especially in smaller communities such as the case study of Shichigahama Town, where almost all survivors managed to rebuild their housing due to strong social and personal connections with the local government staff, who also provided detailed information and advice to each eligible household. Japan may face issues of governance and providing social welfare services in the future, due to a rapidly aging society. However, as a high-income country with established social welfare systems, Japan's government continues to provide a standard level of policy and funding to support disaster survivors.

In contrast, Thailand and India still struggle with development challenges including a large population living in poverty and urban informal settlements. In India's urban city of Bhuj post 2001 Gujarat Earthquake, renters and people living in squatter settlements initially missed out on financial assistance for housing reconstruction. However, due to negotiations by a local CSO, Abhiyan, the local government, later (nearly three years after the earthquake), provided renters with a new relocation site—the GIDC case study site, including land at a subsidized price and housing reconstruction support. Similarly, in Bihar post 2008 Kosi River floods, a few families at odds with the village leader also missed out, reflecting power dynamics at play in the aftermath of disasters. Social class and power dynamics are a double-sided sword as those who had them enjoyed a timely and effective recovery (with some families managing to get more than one house). In Thailand, after the 2004 tsunami, only homeowners were eligible for the initial official government-provided housing reconstruction support. However, in Namkhem with the support of CODI (a Thai government organization), renters and those without land tenure were also able to negotiate housing reconstruction support. Thailand's Namkhem case study and India's GIDC case study therefore represent powerful examples of how recovery policy can be negotiated to expand coverage and to include people who need it most.

### *Resourcefulness (Adaptive Capacities)*

Resourcefulness relates to the resilience of soft assets, that is, capacities of individuals, communities, and institutions to adapt and to make risk-informed decisions, which is of utmost importance. At the individual level, evidence from all case studies suggest some increased awareness of hazard risks, understanding of preparedness (e.g. early warning systems) and what to do in emergencies, as well as an understanding of key multi-hazard safety features in housing. At the community level, communities with pre-existing social bonds and sufficient economic conditions were able to have smooth and speedy housing reconstruction. Moreover, in the long term, such social bonds resulted in improved psycho-social health and well-being. For example, in Japan's small town of Shichigahama, strong community structures of residents and close relationships with local government directly contributed to smoother recovery after the 2011 Great East Japan Earthquake. In contrast, changes in the social fabric of caste-based communities in India's urban relocation site of GIDC in Bhuj weakened social bonds, and in the long term resulted in an additional displacement/ relocation. Conversely, in relocation sites in Namkhem, Thailand, initial weak social connections of neighbors who did not know each other during the first 10 years grew stronger over time, resulting in greater attachment to the community. In addition, the process of community-based reconstruction projects, including self-built houses and the use of community mortgages after the tsunami in Thailand, has strengthened social connections over time. Conversely, as in India, Thailand's poor people and those living in informal settlements struggled with their housing recovery due to lack of social or economic resources and omission from official recovery support. However, with adequate additional support, the poor were able to rebuild housing after the tsunami, which wasn't the case after the 2011 floods, due to inadequate assistance.

At the national government level, an increase in institutional capacities was evident in all three countries. India established a comprehensive Disaster Management Act, the National Disaster Management Authority in 2005, and a dedicated disaster management authority at the state level following the 2001 earthquake. The growth in institutional capacity within the Gujarat government is evident from their leading a reconstruction project with ODHR policy at scale in 2001, as well as playing an advocacy role in subsequent disasters, including in Bihar after the 2008 Kosi River floods, resulting in the evolution of the ODHR program. ODHR policy evolved from 2001 to 2008 in India to incorporate past learnings into subsequent disaster recovery, including all people in housing recovery assistance - not only homeowners, but also landless residents and renters. Conversely, it is unclear if Bihar state government's capacities have improved because recovery after the 2008 Kosi River floods took nearly 10 years to complete (i.e. finished in 2018). However, Indian CSOs' capacities for providing enabling environments to communities during housing reconstruction has also grown, as evidenced by their role changing from mere implementers in 2001 to policy advocates in 2008. Similar to India, in Japan the development of recovery policies also demonstrates the growth of government's institutional capacity. This is demonstrated by recovery policies that evolved between 1995 and 2011, incorporating past learnings into subsequent disaster recovery, extending financial assistance from low-income groups to a broader range of disaster survivors. However, there is limited evidence that this was the case in Thailand, as the recovery after the 2004 mega disaster remained a one-time project, with limited impacts for subsequent (and smaller scale) flooding and earthquake disasters.

## Key Factors with Potential to Catalyze Resilient Recovery Outcomes in Long Term Housing Reconstruction

We observed a few key factors that have helped or hindered resilient recovery outcomes over the long term through cross-country and cross-cultural comparison of six case studies—in India's Bhuj City in Gujarat and Maanganj Town in Bihar, Thailand's Namkhem and Ayutthaya, and Japan's Kobe City and Shichigahama Town in the Tohoku region, as shown in Figure 5.1.

### *1. Leading Role of Government in Crafting an Inclusive and Capacities-Based Recovery Policy*

Although each country had various recovery policies, government (at all levels - national, state, and local) played a leading role in crafting a country-specific recovery policy and overall recovery efforts. In all three countries, creation of a dedicated disaster management authority or disaster recovery authority, who not only crafted country-specific housing reconstruction policies, but could also capture lessons learned, is identified as a key factor in promoting a longer-term view of prioritizing resilience.

Dedicated disaster recovery authorities were established in India in 2001 after the Gujarat Earthquake and prior to the 2008 Kosi River floods in Bihar, and in Japan after the 2011 Great East Japan Earthquake, tsunami, and nuclear accident. India was way ahead in terms of developing a sophisticated institutional setup for disaster recovery, as early as 2001, including multi-scalar and multi-sectoral mechanisms. The 1995 Great Hanshin Awaji Earthquake, affected a smaller regional area, but was Japan's first urban disaster in the 50 years since World War II, and there were fewer precedents to which the government could refer. The scale and complexity of the 2011 tsunami was also unpreceded in Japan and establishing the dedicated Reconstruction Agency was part of the government's response to the necessary scope of recovery. On the contrary, in Thailand, following the unprecedented disaster of the 2004 tsunami, the prime minister took command of recovery in collaboration with six ministries, including the Department of Disaster Prevention and Mitigation (DDPM). However, not having a dedicated disaster authority and rather relying on prime ministers who have shorter political cycles, neither incentivizes longer-term thinking during recovery planning, nor enables capturing learnings to be incorporated in future recovery efforts.

Institutional capacities have been strengthened over time, particularly where dedicated recovery agencies/authorities were created (in India and Japan), due to their ability to incorporate learnings from past disasters into subsequent disaster response and recovery efforts. In India, improved capacity of the Gujarat State Disaster Management Authority (GSDMA) is evidenced by the evolution of owner driven housing reconstruction (ODHR) from 2001 to 2008 following the Kosi River floods, where the government of Bihar extended the financial assistance for housing recovery to include renters and landless residents, to be more inclusive as a result of advocacy of GSDMA. Japan's strengthening of capacities is evident in the evolution of recovery policy from 1995 to 2011, expanded to include a broader variety of options for housing recovery support for disaster-affected municipalities and survivors.

Over time, recovery policies have become more inclusive in all three countries, however India has led the way in bottom-up or what can be termed "capacities-based" policy. In India, despite no precedents to follow for recovery in the aftermath of the 2001 Gujarat Earthquake, the newly formed GSDMA championed an owner-driven housing reconstruction (ODHR) policy at scale, in urban and rural contexts, for the first time, successfully supporting reconstruction of over 100,000 homes in a speedy two years' time, while also ensuring that recovery efforts led to long-term hazard risk mitigation. Upon negotiations with civil society organizations (CSOs), the Gujarat government improvised ODHR policy to include renters in recovery assistance at the case study GIDC relocation site. From 2001 to 2008, ODHR policy was improved to become more

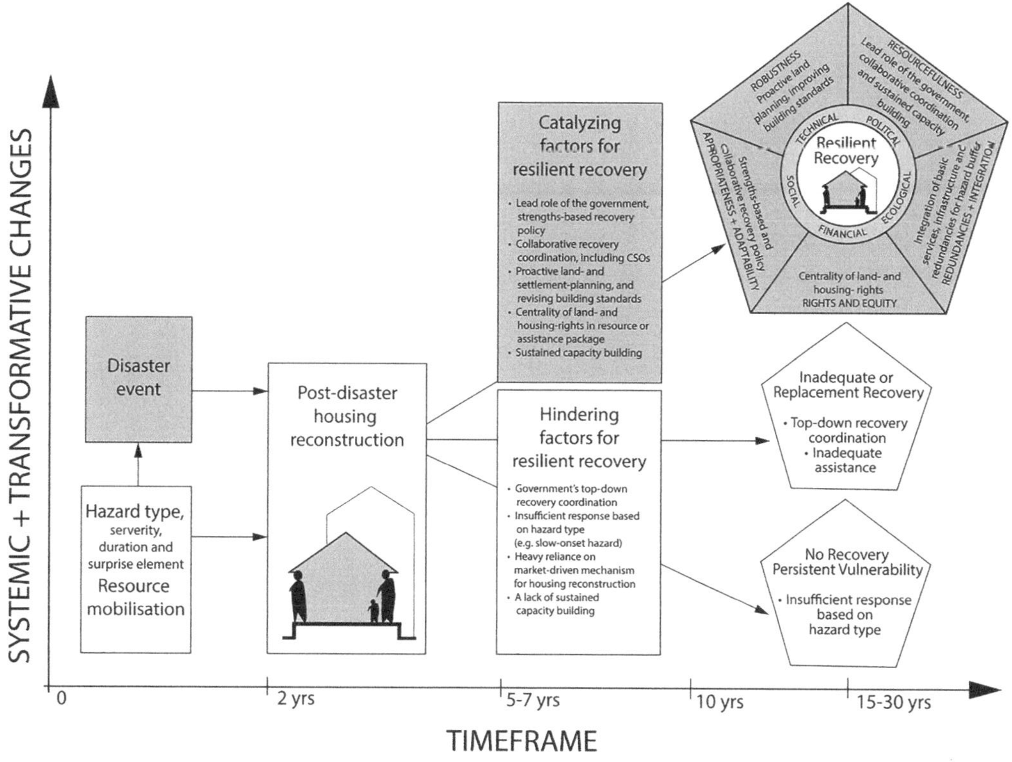


**Fig. 5.1** Model of resilient recovery outcomes in the long term after post-disaster housing reconstruction. (Source: M. Vahanvati with feedback from the authors)

inclusive and better at building upon existing capacities of people in the construction sector and disaster-affected communities. While existing disaster scholarship has already proven ODHR to be the speediest, most dignified, and cost-effective approach to reconstruction (World Bank and GFDRR, 2013; Barenstein & Iyengar, 2010; Barenstein, 2006), evidence from India shows that in the long term (over two decades post-disaster), ODHR programs increased residents' risk awareness, shifted their values, and improved their ability to make risk informed decisions for safer housing extension enhanced by cultural expression through design. ODHR reconstruction has been successful in the long term, mainly due to a combination of assistance, including financial, psychological, social, and technical - and leaving no one behind.

On the contrary, Japan and Thailand, being high-income and middle-income countries, respectively, have adopted a social welfare approach to recovery, including support from both government and donors, whereby mainly financial and technical support was provided (with less emphasis on social support). However, smaller programs carried out with CODI's support in Thailand were closer to an ODHR approach, as they included residents in settlement planning, as well as decisions about extensions, along with self-building of houses.

### *Collaborative Governance and Coordination Mechanisms for Strategic Recovery*

Collaborative governance and coordination mechanisms, evident to varing degrees in all case study sites, ensured incorporation of diverse stakeholders' values and knowledge in decision making, removing conflicts or competing interests, in early stages of recovery. Serving as a key factor, collaborative coordination mechanisms facilitate bringing together people, resources, processes, and governance—through decentralized, localized, actionable, one-stop-shop recovery hubs in India (e.g. *Setu Kendras or Kosi Setu Kendras*) or participatory *machizukuri* processes in Japan—ensuring a contextually appropriate and strategic recovery, or resilient recovery that merits replication.

Collaborative coordination—multi-scalar and multi-sectorial—was evident in India's Bihar after the 2008 Kosi River floods in the form of *Kosi Setu Kendras* (KSKs), and in Japan after the 1995 Hanshin Awaji Earthquake in the form of *machizukuri* (community planning) committees and processes, both of which played a vital role in community participation, raising concerns while empowering residents.

In India, coordination was top-down and hierarchical in Gujarat after the 2001 earthquake, except for a consortium of CSOs—Abhiyan, who experimented with decentralized coordination evident in *Setu Kendras*, established as one-stop-shops in Kachchh. From 2001 to 2008, India took a big leap from centralized to decentralized coordination, informed by learnings from experience in Gujarat. The informal setup of Setu Kendra was formalized and incorporated in Bihar state government's institutional body to form *Kosi Setu Kendras* (KSKs), which literally translates to "bridging-centers." Each KSK was comprised of social workers, lawyers, finance professionals, engineers, and information technology personnel, serving 4000 families. KSKs were set up in Bihar as one-stop-shops or shelter hubs to serve as a conduit of communication between government and communities, raising concerns of residents in a manner that gets heard, and providing essential information to communities to make informed decisions.

Similarly, in Japan, *machizukuri*, which literally translates as town-making, represents a participatory process that goes beyond working towards community building and community improvements. As an established idea in community development, it took on a key role in recovery after the 1995 earthquake in Kobe; *machizukuri* has become part of mainstream community planning and refers to a variety of community planning processes and projects, with actual levels of participation varying greatly (as with the term "community participation" globally). After the 2011 earthquake and tsunami, the roles and importance of *machizukuri* in the recovery process and projects also varied across multiple communities in the many tsunami-affected municipalities. In the small town of Shichigahama, post-2011 GEJE, the local government played an important role of coordination, based on a pre-existing relationship of trust with their community.

In Thailand, the Community Organizations Development Institute (CODI)—a Thai government institution—existed to serve low-income groups or those living in informal settlements. While CODI in Thailand and KSKs in India were similar in terms of being part of a government institution but working with communities at the grassroots level, before the tsunami CODI had only served those living in informal settlements. KSKs in India, on the other hand, served all disaster survivors and continue to serve communities in non-disaster times, to date. In Japan, *machizukuri* represents an approach to participatory planning for disaster as well as non-disaster times. In all the case studies, these hubs, facilitators, or approaches (e.g. KSKs, CODI or *machizukuri*, respectively) created effective channels of communication and negotiation between government and communities, generating positive outcomes.

*Weakness of Top-Down Coordination*

In contrast, government' top-down coordination negatively impacted resilient recovery outcomes. Top-down and centralized governance and coordination mechanisms were evident in India (after the 2001 Gujarat Earthquake), in Japan (after the 1995 earthquake and 2011 GEJE), and in Thailand (after the 2004 Indian Ocean tsunami and 2011 floods), but with minor differences. In Thailand, disaster type and scale influenced the form of governance in response and recovery, with the massive and rapid-onset 2004 Indian Ocean Tsunami disaster led by the national government in a command-and-control style, while response after the slow-onset disaster 2011 floods with fewer human casualties was led by the state government and implemented by donors. In Japan, recovery efforts are almost entirely managed by government, with the national government providing funding and local government administering it, with very limited roles of CSOs. In Japan, recovery planning was done in a two-tiered manner in 1995, wherein city recovery plans were developed by the government as a first step, followed by revisions proposed with the involvement of local community planning organizations and residents. Although still limited, the roles of non-profit organizations (NPOs) and volunteer groups in supporting disaster survivors in Japan continues to grow. While top-down approaches lead to speedy recovery, they have limited ability to build people's capacities within 2 to 3 years. Japan's top-down recovery approach has built government's capacities over time; however, that cannot be said for all survivors, especially low-income groups, who had limited choices in decision making for their housing. In Japan, it is uncertain whether capacities of high-income residents, who had freedom to decide about their housing reconstruction, increased or not. Conversely, residents who had the financial resources to rebuild on their own could also choose not to participate in community-based activities.

### *The Enabling Role of Civil Society Organizations (CSOs) in Developing Countries*

Civil society organizations (CSOs), non-government organizations (NGOs), or non-profit organizations (NPOs) played a vital role in low- or mid-income countries of India and Thailand in advocating for the rights of marginal groups, ensuring government's accountability in recovery programs and implementing community-centered, inclusive recovery efforts. Active roles of CSOs as well as government–CSO partnerships were evident in both Indian cases and in Thailand.

In India, after the 2001 Gujarat Earthquake, through public–private partnerships, government allowed CSOs to adopt communities to facilitate housing reconstruction implementation. CSOs played a major role in translating ODHR policies into practice, through skills training of construction workers, awareness raising among residents, and providing hand-holding support to residents with construction supervision for quality checks. Through their active role and experience in disaster recovery processes, CSOs became experts, and their roles evolved from 2001 to 2008, from being mere implementers of housing reconstruction projects in Gujarat to becoming ODHR policy advocates and government capacity builders in Bihar. The mature and contextually responsive ODHR program tailored in Bihar by CSOs and the government of Bihar collaboratively, have the potential to have larger-scale positive impact; the active role of CSOs as recovery policy advocates merits continuation.

In Thailand, NGOs' work in disaster recovery was very important because they had more flexibility for funding, could work faster, and were closer to community networks and residents. After the 2004 tsunami, many NGOs came to southern Thailand to support recovery, not only by providing funding but sometimes also providing designs for post-disaster houses, materials, and planning. Not only was the quality of houses provided by NGOs better compared to houses provided by the government and constructed by the military, in addition to support for housing reconstruction, NGOs continued to support residents' recovery processes, such as by funding house extensions, providing education, and supporting safety practices for the local community. After the 2011 flood, the Red Cross and other NGOs also supported residents by providing food and water. Although many donations from across the country and internationally were used for infrastructure reconstruction, for household recovery, most renovations were carried out by individual residents' decisions and actions.

In Japan, the role of CSOs is relatively small, with the government carrying out most policy and programs. The 1995 earthquake became known as "year 0 of volunteering" in Japan; non-profit and volunteer organizations have been passionately supporting affected communities after every disaster since then in growing numbers. While strongly involved in various activities supporting survivors, including community building in temporary and permanent housing, non-profit organizations still play a very limited role in housing reconstruction.

### *Proactive Land and Settlement Planning and Improving Building Regulation for Strengthening Resilience*

#### *Land Use Regulation Changes*

Land use planning changes post-disaster allowed large-scale reduction of exposure and vulnerabilities. Of the three countries, Japan is the only one where government engaged in land use planning changes, based on hazard risk mapping, rather than focusing on individual buildings, which is worth learning from and replicating. Land readjustment projects (including property reorganization) were carried out across many affected areas of Kobe City after the 1995 earthquake. Affected by not only an earthquake but also an unprecedented tsunami, and in consideration of future tsunami risk, risk-based land use planning took on a larger role after the 2011 GEJE. Major land use changes implemented as part of municipal recovery plans required residents to be moved out of high tsunami-risk areas onto safer higher lands through projects that include government programs to acquire former land and prepare new residential lots in safter areas.

Settlement-scale land readjustment and property reorganizing for road widening for ease of emergency vehicle access reducing hazard risk was evident both in Japan's Kobe City and in India's Bhuj City's urban core renewal post 2001 earthquake. This was the first time the Bhuj Area Development Authority (BHADA) had embarked on an urban renewal project. Lacking prior experience and with an outdated development plan, they engaged a consultant for the urban renewal of Bhuj, which caused over 40% of residents to be relocated out of the walled city. The Government of Bhuj supported people's relocation by providing relocation sites on the outskirts of Bhuj. One of Bhuj's relocation sites, GIDC, also had a settlement layout, perhaps because it was a green-field site, but such settlement-scale design allowed integration of open spaces for community gatherings, services, and infrastructure. Internationally, a settlement-based, area-based, or place-based approach is increasingly promoted due to an understanding of interdependence between physical infrastructure (such as community gathering spaces or religious places or sports facilities) and social dynamics. In both India's Bhuj City and Japan's Kobe City, government focused on urban infrastructure development for the whole community, which also led to minor land use planning changes.

In Thailand after the 2004 tsunami, although hazard risk information was disseminated based on risk mapping, residents were not required to move away from hazardous areas in order to receive official support to rebuild on site. Some relocation settlements were built in safer areas with donor support.

The following three levels of land use planning regulation changes were evident in all three countries' case studies, as: (1) land use changes deeming a site unsafe for inhabitation (requiring major relocations), (2) land adjustment on site, causing minor changes to existing land use, and (3) settlement layout on green-field sites, with recently changed land use to accommodate relocated residents. Land use changes are sometimes unavoidable, especially when hazard risk is high. However, it is not clear if relocations in India's post-earthquake case of Bhuj or Thailand's post-tsunami case were informed by future hazard risk mapping. Changes in planning regulations, by land use changes, overlays, or planning scheme amendments, are vital for reducing exposure; however, they should be informed by scientific data on future hazard risk mappings.

*Improved Building Standards Offer Buffer Capacity for Housing to Withstand Future Hazards*

Buildings codes were revised after the disasters in both Indian cases (in 2001 and 2008). India was the only country where traditional and green construction technologies like stabilized earth, bamboo, or stone construction were integrated into new building codes. However, in the case of India, despite some of the best building regulations and legal frameworks published by the Bureau of Indian Standards, enforcement remains lax (Arya et al., 2014).

Japan had already revised seismic requirements in the building code in 1981, so 1995 and 2011 disasters did not impact building code revisions. As enforcement and compliance rates are high in Japan, as is construction quality overall, all buildings built since 1981 comply with these codes. In

Thailand, although building codes were not revised after the 2011 flood or 2004 tsunami, risk maps and risk zones were created for tsunami hazard areas.

### *Centrality of Resources for People's Ability to Recover*

Resources—human, financial, social, material, and land—are key factors determining disaster survivors' ability to rebuild housing and to recover as a way out of vulnerability in the long term. While financial assistance is of utmost importance, we found technical know-how and psycho-social support for survivors to navigate recovery amid disaster trauma to be equally important. While it is hard to precisely specify the amount or type of combination of resources necessary for recovery, as it depends on multiple variables, evidence from all three countries demonstrate that income levels often determine the types and amounts of resource assistance offered by the government to residents for housing reconstruction.

#### *Resource Amount and Combination/Type Matters for the Most Vulnerable*

In India, a low-middle-income country where insurance or market-driven mechanisms are not prevalent, recovery relied on the state governments of Gujarat and Bihar, who borrowed money from the World Bank after the 2001 earthquake and 2008 floods, respectively. Survivors in both case studies received financial assistance from the government as well as CSOs. Survivors who had to relocate or were renting (in Bhuj) or were landless (in Bihar) also received additional financial support from the government to purchase land at a subsidized rate. The government's financial package in Gujarat post 2001 earthquake was based on damage assessment, urban or rural location, and type of house. In Gujarat, the government's financial package was supplemented by CSOs or NGOs, whose assistance varied from provision of fully built houses, to mere technical assistance, to a combination of financial and/or material, social, and technical support. However, assistance packages changed from 2001 to 2008. In Bihar, post 2008 floods, uniform assistance packages—including, financial, social, and technical support, through KSKs, were devised by the state government of Bihar for all residents. The assistance was uniform, mainly due to all flood-affected people living in rural areas with similar house types. Such an assisted yet resident-led recovery process is coined by Vahanvati et al. (2023) as "assisted self-recovery". Similarly, in Thailand, an upper middle-income country, a combination of government and CSO support was provided to survivors after the 2004 Indian Ocean tsunami, leading to a range of options from fully built housing to house repair to self-help housing (with only technical assistance). A combination of assistance packages helped survivors navigate the complexities of the building and construction industry in a timely manner. The Indian government's creative method of resources allocation as a combination of government and CSOs, and types of assistance packages (including financial, technical, social and psychological) is worth replicating.

Lack of or inequitable access to resources, such as funding or land, is a major factor hindering residents' resilience. Equity issues were evident in both case studies in India and Thailand, although less so in Japan. In India's Gujarat after the 2001 earthquake, the state government's non-uniform financial package was critiqued for being biased towards the rich and those living in *pucca* (or permanent) houses, who received more aid, while renters and people living in informal settlements who needed most help missed out on any form of recovery assistance. To date, renters living in Bhuj's GIDC area have not secured land rights and are resorting to informal transfer of property ownership and land titles. In Bihar after the 2008 floods, corruption among village-level government and social power dynamics undermined equitable distribution of resources and thus recovery of all. At the government scale, the state governments of Gujarat and Bihar feel the burden of loans and interest repayment to the World Bank. Similarly, in Thailand's Namkhem after the 2004 Indian Ocean Tsunami and in Ayutthaya after the 2011 floods, informal residents missed out on recovery assistance. The ongoing struggle of informal settlement dwellers for secure land tenure and safe housing remains a major factor compounding their risk.

In contrast, in Japan—a high-income country—the government took a more hands-off approach towards the reconstruction of private housing in 1995, providing almost no assistance to homeowners, who needed to rebuild their private homes through the private housing market (i.e. bank loan or savings, as disaster insurance played a relatively small role in reconstruction support). Government policies for supporting the reconstruction of private housing changed several years after the Kobe Earthquake, and for disasters since 1998, including after the 2011 earthquake and tsunami, owners of completely damaged houses can receive up to 3,000,000 yen (20,780 USD in 2024), as stipulated in the 1998 Act on Support for Reconstructing Livelihoods of Disaster Victims. After the 2011 tsunami, along with the creation of the national Reconstruction Agency, the Japanese national government created a recovery program (with subsidies termed *kofukin*) that included a menu of reconstruction projects municipalities could include in their towns' recovery plans. For housing recovery, these projects included Collective Relocation for Disaster Mitigation, which is a two-part program that can (1) purchase residents' former land designated as being in a hazard risk area and (2) provide residential lots (including basic services and infrastructure) on safer higher land for relocation. The Japanese government continued to provide disaster recovery public housing for disaster survivors in 1995 and 2011, as in-kind housing, built by the government, with income-sensitive government-subsidized rents. Although tsunami damage washed away many houses completely, for those with partial damage, the government also provided some support programs for housing repair. Similarly, the Thai government also adopted a social welfare approach for low-income or the most vulnerable, while promoting self-help housing for the rest.

These examples suggest that countries with higher levels of development and income may rely too much on market-driven mechanisms for housing recovery, leaving people to recover using their personal economic and abilities, conversely missing opportunities to build residents' capacities. Money is the starting point for recovery, but the amount of financial allocation and survivors' control over the allocated money is what determines recovery priorities.

*Equity issues in resource allocation*

In India, after the 2001 earthquake as well as 2008 floods, it is commendable that the state government prioritized gender equity, mandating joint titles for property between husbands and wives, as well as joint bank accounts into which the government transferred money, in three tranches. While transferring money into residents' banks helped maintain transparency, it also led to further liberalizing of the economy. While the funding dispensation was timely in Gujarat, for homeowners (not renters), that was not the case in Bihar after the 2008 floods; piloting before the implementation of the main policy caused delays of over two years before aid could reach most affected people. In addition, the practice of disbursing financial aid money to residents in three tranches, contingent upon rebuilt houses' compliance with building standards, encouraged robust building practices and merits replication.

People who owned their houses and land had more opportunities for housing recovery after the 2004 tsunami in Thailand. Their options for reconstruction included: (1) receiving a house provided by the government (and partial donor support) and built by the military; (2) receiving a house provided by a donor organization, built by construction professionals and provided in a settlement in a new relocation area; or (3) receiving 50% of the support funds (compared to option 1), to rebuild on their own land by themselves. Each choice included varying levels of community participation in the recovery process. Most tsunami survivors in Namkhem worked in fishery industries, and many lived in a slum near the shore. Without land ownership before the tsunami, they did not qualify for the standard housing recovery support provided by the government to rebuild housing for tsunami-affected residents on their own land. With the support of CODI, along with other international NGOs, the residents participated in many processes. Starting from gathering in community networks, they worked out which location was safe and available for planning their new housing settlement. CODI also helped finance the reconstruction through a community mortgage; every month, residents made payments and 15 years later they received land rights and ownership. Even with the ownership they had before they received the land rights, they could sell the house to others and adapt their own houses.

In Japan, after the 1995 earthquake in Kobe, since the government provided fully built, multi-family apartments to eligible low-income residents via subsidized rents in the form of disaster recovery public housing, their recovery was relatively speedy, mostly within 3–5 years. Similar approaches were also adopted after the 2011 GEJE. In 2011, disaster recovery public housing included additional housing types, not only apartment-style multi-family buildings but also single-family detached housing units, built close to (or relocated to nearby high land from) former villages. While the 1995 earthquake recovery was relatively speedy (approximately 3 years) for some survivors, relocation projects after the 2011 earthquake and tsunami required more time for land use planning and land preparation. In addition, the nuclear meltdown at the Fukushima Daiichi Nuclear Power Plant resulted in long-term evacuation and uncertain recovery for affected survivors in Fukushima. These examples demonstrate that land use planning changes, essential after disasters, and hazard type often hinder timely recovery.

Floods, as slow-onset disasters, lack visibility from governments or donors, and therefore lack strategic recovery programs or funding assistance necessary for flood-affected people's recovery. We examined two cases of post-flood reconstruction, one in India after the 2008 Kosi River floods in Bihar and another one in Thailand, after the 2011 Chaopraya River floods in Ayutthaya. In both cases, despite residents' growing awareness of flood risk, their reliance on agriculture-based livelihood meant they were not willing to move away from their homes. In Ayutthaya, residents were aware of the need to raise houses' floors to improve robustness and avert future damage to their houses and belongings during subsequent floods. However, lack of adequate financial assistance was the prime barrier for Ayutthaya residents to improve their housing. In contrast, India's Bihar state government engaged in a strategic housing recovery program, assisting approximately 60,000 families, who each received 1930 USD to build a core house with elevated floors, an attic, and multi-hazard safety features for floods, earthquakes, and high winds, improving the houses' robustness to withstand future hazards. However, such strategic recovery programs are rare after floods.

*Shortfalls of market-driven recovery programs.*
The three-country six case study evaluation revealed that higher income or development levels of a country and megadisasters (e.g. tsunami) may result in a reliance on neoliberal, market-driven means for housing recovery. Housing reconstruction in the high-income country of Japan can be understood as following a two-track recovery since the 1995 Kobe Earthquake. In one track, private housing reconstruction relied on market-driven mechanisms for housing recovery, with people's recovery heavily depending on their personal income or economic conditions. Another track was the provision of disaster recovery public housing by the government, through the public sector.

*Social issues emerge from post-disaster recovery efforts*
Housing in developed countries is strongly connected to the economic development of the nation. As housing recovery becomes stratified based on survivors' economic resources, it can ignore social, cultural, or ecological aspects of the society within which housing is embedded. As is the case in most disasters around the world, homeowners with the economic means to rebuild on their own do so, as was the case after both disasters in Japan, in 1995 and 2011, and is increasingly the case in other countries. In Japan, policy changes have resulted in increased support for private housing reconstruction to a degree, and after the 2011 Earthquake and tsunami,

government recovery programs funded collective relocation projects as a support mechanism for residents' reconstruction. As in other countries, Japan's low-income survivors had and continue to have fewer housing recovery options, relying on disaster recovery public housing built and provided by the government. After the 1995 earthquake, allocation by lottery of public housing units, most located far from former neighborhoods, disregarded social fabric and negatively impacted community ties. Along with limited choices about houses' spatial design or location, the result was social isolation for many low-income residents, contributing to an epidemic of solitary deaths.

After the 1995 earthquake, capacity-building efforts for stronger communities and disaster risk reduction occurred in parallel to the reconstruction of housing. As part of the participatory design process for community recovery, neighbors in active areas developed stronger bonds. Even before the 1995 earthquake, Japan had and continues to have high levels of structural resilience in buildings and infrastructure. There have certainly been advances in the understanding of the importance of capacity building and social capital. However, the siloed nature of reconstruction and housing projects, and emphasis on engineering and infrastructure, limits opportunities for holistic integration of social, community, and disaster resilience into recovery projects.

While Thailand continues to experience a rapid rise in development status, following the income-dependent and economic-driven policies of other developed countries, such a rise may lead to reduced social aspects, in turn weakening community resilience. Government policies related to river water management and floods in Ayutthaya in 2011 also prioritized the economic activities of the private sector directly, including the decision to intentionally flood certain residential areas to protect the industrial zone and private corporations.

### *Sustained Capacity Building as a Dynamic Process for Strengthening Resilience in Housing Systems*

Our findings underscore the significance of sustained capacity building and continuous learning as a dynamic process, essential for advancing resilience in the planning, architecture, and construction sectors as well as within government institutions, housing and disaster stakeholders, and communities. The long-term effectiveness of housing recovery programs and practices hinges on continuous education, training, monitoring, learning, and improvements based on those learnings. A lack of such sustained capacity building within government institutions, the private sector, and residents, emerged as a major barrier to maintaining and advancing housing resilience.

Especially in India and Thailand, the case studies highlight a need for expansion of pre-disaster hazard risk mapping for governments to be able to continually update land use planning, improve preparedness (e.g. early warning systems), raise community awareness, and conduct training in the construction sector, including on enforcement of land use and building regulations. A huge gap persists between all the efforts put into robust housing reconstruction post-disaster and ongoing housing construction as a development activity when enforcement and compliance of building standards become lax. Within the building sector, it was evident in the Indian case studies that new skills learned during recovery were getting lost over time. For example, in India's Bihar after the 2008 Kosi River floods, without adequate and continuous training to convert learned skills into certification or embedding them in livelihood means, safe construction did not become part of housing culture. At the residents' level, people have started forgetting about hazard risks or safety features in housing due to the long interval since the last disaster was experienced, demonstrating a loss in risk-informed decision-making ability. For example, green technologies such as bamboo construction or solar lighting, introduced in Bihar post-2008 floods, and green water sewage recycling and garbage conversion to vermicompost developed at the GIDC site post-2001 Gujarat Earthquake, were in dilapidated condition due to residents' lack of awareness about or lack of maintenance skills. Strengthening resilience in housing systems require the longevity and safety of reconstructed housing and beyond, that is, sustained and vigilant governance for embedding such practices in development activities. One key barrier to sustaining institutional capacity growth, at least in India, is a lack of hazard risk mapping to understand risk and inform land use planning changes. Having said that, the government of India has invested in developing this capability.

Continuous learning and improvement based on previous experiences underscores the need for a dynamic approach to resilience building in the Asian context.

## Conclusion

Based on cross-country and cross-cultural comparative analysis of resilience outcomes in the housing systems over the long-term after six post-disaster housing reconstruction projects, after two major disasters in each of three Asian countries—India, Thailand, and Japan—we identified five key factors that helped or hindered systemic changes for strengthening resilience, which is imperative for society to thrive in a changing climate. Informed by these five key factors, we propose a few recommendations with the hope of informing future recovery interventions, domestically, regionally within the Asia-Pacific, and internationally, for the resilience imperative of our times. Linked to the five key factors that contributed to building resilience and capacities, among all (residents, CSOs, and governments), these recommendations include the following.

(i) **A leading role of government and dedicated disaster management and recovery authorities for crafting an inclusive and strength-based recovery policy**. This was a key factor in helping build capacities at all levels (government, planning, design, and construction sector and residents). Additionally, having a dedicated recovery authority helps build institutional capacity over time, by monitoring and continuous learning to inform future interventions. India has a dedicated government authority and championed an ODHR policy. Japan has a dedicated authority, while Thailand lacked such an authority but adopted community-driven recovery approaches in some cases.

(ii) **Collaborative governance and recovery coordination**, including government, CSOs, and the private sector, in the decision-making process. This played a vital role in developing strategic goals during community-based housing recovery, because multi-scalar and multi-sectoral collaboration allowed the incorporation of diverse stakeholders' values and knowledge in decision making. Moreover, such collaborative processes allowed for negotiations between the government and civil society organizations (CSOs), who advocated for the rights of the most vulnerable groups, particularly in India and Thailand. Mainly evident in the contributions and impacts of *Setu Kendra* and *Kosi Setu Kendra* in India, and *machizukuri* in Japan, as community hubs, and these systems merit replication, as do the approaches adopted by CODI in Thailand.

(iii) **Proactive land- and settlement-planning and improving building standards for strengthening resilience.** This is a key factor for creating buffer capacity or spatial-technical resilience. Only the post-disaster recovery phase allows for large-scale transformative changes at city- or settlement-scale to reshape, rezone, or regenerate land and to create built-in redundancies to reduce exposure to future hazards. Such large-scale transformations were mainly evident in Japan and in some cases in India (Bhuj) and Thailand (Namkhem).

(iv) **Centrality of rights for resilience building, in tailoring resources/assistance packages that are also contextually appropriate**. Resource combination, type, amount, timeliness, and distribution affect people's ability to recover, especially in low- and middle-income countries. India and Thailand continue to have challenges with inadequate land- and financial assistance, but lead the way in socio-technical assistance, whereas Japan leads in technical and financial assistance, and sustaining capacity building (or soft assets' resilience) at all levels and various sectors to continually improve adaptive capacities, that is, resilience as a dynamic practice.

(v) **Sustained capacity building as a dynamic process for advancing resilience in housing systems.** Continual awareness raising among residents, and skills training of the local construction sector (to keep up with the improving building standards), and continual learning by government authorities, relevant stakeholders, and CSOs are the most significant factors to foster the continuation of resilience strengthening as a dynamic practice. This is the most significant aspect of ensuring capacities built during recovery are transformed into capacity development, that is, during ongoing development activities. Japan has established a culture of such continual learning at all levels; India's government and CSOs are continually building their capacities through collaborations but such capacities are not maintained within the planning or building sectors or at the community level. Thailand lags in this area, potentially due to a lack of dedicated government authority for disaster recovery. Potentially, India and Thailand can learn from Japan about attaining a high compliance culture in the planning and building sector.

We also identified several key factors that hindered resilience outcomes or systemic changes including: (i) negative impacts of government's top-down coordination mechanisms, which left a lasting legacy of psycho-social and well-being challenges for survivors, (ii) inadequate financial assistance following slow-onset hazards (e.g. floods) leading to persistent vulnerability, (iii) the shortfalls of market-driven recovery programs, and (iv) a lack of sustained capacity building beyond housing reconstruction completion into ongoing development activities, for futureproofing and adaptability of housing systems including its residents.

We hope that learnings from these long-term systemic changes identified from across three Asian countries, with varying development levels, geoclimatic context and cultural backgrounds, would inform future recovery interventions, domestically, regionally, and internationally, to attain the twin imperatives of our times - rights for all and resilience in housing systems over the long term.

## References

Arya, A., Boen, T.. & Ishiyama, Y. (2014). *Guidelines for Earthquake Resistant Non-Engineered Construction* (June ed.). United Nations Educational, Scientific and Cultural Organization.

Barenstein, J. D. (2006). *Housing Reconstruction in Post-earthquake Gujarat: A Comparative Analysis.* Network Paper [Online], 54. Retrieved May 2019, from https://odihpn.org/wp-content/uploads/2006/04/networkpaper054.pdf

Barenstein, J. D., & Iyengar, S. (2010). India: From a Culture of Housing to a Philosophy of Reconstruction. In M. Lyons, T. Schilderman, & C. Boano (Eds.), *Building Back Better: Delivering People-Centred Housing Reconstruction at Scale*. Practical Action.

Comerio, M. C. (2014). Disaster Recovery and Community Renewal: Housing Approaches. *Cityscape (Washington, D.C.), 16*(2), 51–68.

Desai, R., & DesaI, R. (2011). Citizen Architects in India. In M. J. Aquilino (Ed.), *Beyond Shelter: Architecture for Crisis.* Thames & Hudson.

Johnson, L. A., & Olshansky, R. B. (2017). *After Great Disasters: An In-depth Analysis of How Six Countries Managed Community Recovery.* Lincoln Institute of Land Policy.

Vahanvati, M. (2015). *Reviving People's Trust in Bamboo Technology: A Case-Study of Orlaka Settlement Reconstruction in Bihar, after the 2008 Kosi floods.* In 2015 World Bamboo Congress, 17–21 Sept, 215 2015 Damyang, South Korea. WBC2015.

Vahanvati, M., McEvoy, D. & Iyer-Raniga, U. (2023). Inclusive and Resilient Shelter Guide: Accounting for the Needs of Informal Settlements in Solomon Islands. *International Journal of Disaster Resilience in the Built Environment, 14*, 577–590. Retrived Dec 2023, from https://doi.org/10.1108/IJDRBE-10-2022-0098

World Bank, & GFDRR. (2013). *Enhancing Safety and Improving Living Conditions for 100,000 Vulnerable Families: Building a Culture of Resilient Reconstruction in India.* World Bank.

# Index




M. Vahanvati et al., *Resilient Recovery from Disasters*,
https://doi.org/10.1007/978-981-97-8047-1

**I**

J



K

## M



## N

## T